Plant Growth Regulators

Chemical Activity, Plant Responses, and Economic Potential

Charles A. Stutte, EDITOR

University of Arkansas

A symposium cosponsored by the
Division of Pesticide Chemistry
and the Plant Growth Regulator
Working Group at the 170th
Meeting of the American
Chemical Society, Chicago, Ill.,
August 27–29, 1975.

ADVANCES IN CHEMISTRY SERIES **159**

AMERICAN CHEMICAL SOCIETY

WASHINGTON, D. C.   1977

Library of Congress CIP Data

Plant growth regulators

(Advances in chemistry series; 159 ISSN 0065-2393)
Includes bibliographical references and index.

1. Plant regulators—Addresses, essays, lectures.
I. Stutte, Charles A., 1933–   . II. Series: Advances
in chemistry series; 159.

QD1.A355 no. 159 [SB128] 540'.8s [631.8] 77-7934
ISBN 0-8412-0344-X

Advances in Chemistry Series

Robert F. Gould, *Editor*

FOREWORD

ADVANCES IN CHEMISTRY SERIES was founded in 1949 by the American Chemical Society as an outlet for symposia and collections of data in special areas of topical interest that could not be accommodated in the Society's journals. It provides a medium for symposia that would otherwise be fragmented, their papers distributed among several journals or not published at all. Papers are refereed critically according to ACS editorial standards and receive the careful attention and processing characteristic of ACS publications. Papers published in ADVANCES IN CHEMISTRY SERIES are original contributions not published elsewhere in whole or major part and include reports of research as well as reviews since symposia may embrace both types of presentation.

CONTENTS

PREFACE

Manipulation of crop production processes with chemicals may be one of the most important advances to be achieved in agriculture. Past successes in regulating plant development, increasing interest in expanding the world's food production, along with the need for more efficient use of available energy have stimulated renewed interest in controlling plant growth processes to obtain more efficient productivity.

Since our understanding of plant physiological processes is incomplete, much research remains to be done before the potential uses of plant regulating substances to obtain a desired response can be realized. However, as more information on plant responses to chemicals does become available and an economic potential is established, development of chemicals for practical use will be accelerated. In the past decade we have seen rapid growth in the use of regulants that have proven profitable to the growers and manufacturers and beneficial to the consuming public. It is the purpose of the papers in this publication to discuss some of the successes and some of the problems involved in plant growth regulator research and the subsequent development of the products for use by growers. As more interaction between scientists in industries and universities occurs, growth regulant technology should advance more rapidly. The papers of this series are of value to individuals interested in information that will aid in determining the potential for using chemicals to alter specific physiological processes. These altered processes should provide a desired plant response that will prove beneficial to mankind.

University of Arkansas   CHARLES A. STUTTE
Fayetteville, Ark. 72701

Chemical Activity and Plant Responses

Considerations in Searching for New Plant Growth Regulators

ERNEST G. JAWORSKI

Monsanto Agricultural Products Co., St. Louis, Mo. 63166

Interest is strong in the study of plant growth regulators as reflected by the increasing number of research publications and the fact that at least 29 major companies are now actively engaged in this type of research. Nevertheless, advances to date have been limited by complications confounding straightforward elucidation of the relationships between chemical structure and plant response. Plant growth regulation is expressed by an integrated series of reactions and interactions. The ability to influence these reactions toward some desired goal represents a more complex situation than one where a simple lethal effect (herbicide) is desired.

Structure–activity data for plant regulators are not very abundant in comparison with similar information on pesticides. This stems partly from the relative novelty of plant regulators and partly from the difficulty in obtaining biological test data. While research in plant regulation has been active for scores of years, identification of specific regulators—more precisely, plant hormones—has been generated within the past 35–40 yrs.

Auxins, and more specifically indoleacetic acid, have probably received the greatest attention, having been the first type of plant hormone to be characterized. Research in this area led to such auxin analogs as naphthaleneacetic acid and the phenoxyacetic acids and to their commercial application in fruit set, fruit thinning, and other regulatory (non-herbicidal) uses. The subsequent discoveries of other natural plant hormones like gibberellins, cytokinins, and abscisic acid led to a similar evolution of analogs although the commercial applications were, with the possible exception of gibberellic acid, on a much more modest scale than the auxins. Ethylene and ethylene-generating chemicals such as ethephon have probably received the most intensive scrutiny in recent times where structure–activity relations are concerned. Even so, relatively little published data are available regarding structure and biological activities for all hormone classes presently recognized.

The discovery of naturally occurring hormones has also generated considerable interest in the synthesis of chemicals that would either block the endogenous synthesis of the hormone or interfere with its transport from the site of formation to the target site of action. Such approaches have resulted in the development of a number of commercial or near commercial chemicals such as Alar, which is used to promote increased flowering and fruit set. Thus the search for structures analogous to known plant hormones or their precursors has served as a fundamental approach to finding plant regulators over the past three decades.

Knowledge of the pathways of hormone biosynthesis and the mechanisms for their regulation would be especially useful in developing chemical modifiers that could alter hormone levels in desired places in the plant and during appropriate temporal stages of its development. For example, it is conceivable that fruit retention in cotton and soybeans would enhance yield. Enhancement of auxin transport from leaves to newly developing reproductive organs could result in such retention by preventing abscission, as suggested by Addicott some time ago.

Possibly one of the best cases that exemplify the difficulties in defining structure—and what we perhaps should now refer to as commercial activity or agronomically significant activity—is one that could be called the TIBA story. TIBA, 2,3,5-triiodobenzoic acid, was felt to have great potential in increasing soybean yields. Numerous analogs and homologs were studied, and this particular compound appeared to be the best—at least under the test conditions used. Field results, unfortunately, were erratic and ranged from highly positive to highly negative effects on yield. Large differences in the responses of different cultivars were noted, and strong interactions with management and cultural practices as well as with environmental effects, especially early season moisture levels, were all encountered in seven years of field research with TIBA. One wonders whether some other derivative, though perhaps of lesser activity under the standard test conditions, might have been more effective under varying field conditions.

The TIBA story highlights at least two major considerations that need more attention in future structure–activity evaluations of plant regulators.

(1) We must recognize that plants, unlike animals, cannot move about. They are literally stuck with whatever environment they find themselves in. Natural selection over tens of millions of years has therefore resulted in the remarkable adaptability we see in plants today. Their ability to adjust to adverse environmental shifts, however, could also result in internal changes that either work indirectly to counteract a desired plant regulator effect or directly to correct for the elicited response. In either event, the plant regulator effect might then be short-lived. More sophisticated determinations of endogenous biochemical and

physiological activities will probably be required to develop a greater understanding of our plant regulator candidate.

(2) The environmentally elicited responses within the plant will have to be superimposed on plant regulator studies under controlled conditions if useful structure–activity correlations are to be made.

Plant regulators must be examined from the point of view of the plant. There is interest in improving photosynthesis and yield by blocking photorespiration—a process considered by some research workers to be unnecessary and even wasteful for the plant. In soybeans, we have found that the leaves can increase their photosynthesis rates by 50% simply in response to the development of the bean and the bean pod. In other words, the plant contains endogenous mechanisms for enhancing photosynthesis in response to a sink demand by a reproductive organ. This example of inherent endogenous response capability in the plant illustrates how critical the timing of an application of a plant regulator could be in seeking to find a photorespiratory inhibitor to enhance photosynthesis.

Thus the stage of plant development, the timing of chemical treatment, and the environment all require consideration in structure–activity studies. When we superimpose genotypic and phenotypic variations upon this complex of variables, it is apparent that plant regulator research and development is considerably more complex and requires more rigorous and critical evaluation criteria than has been customary in pesticide research. However, attention to such detail should promote discovery of useful chemicals which will play a highly significant role in our future efforts to improve food quality and increase crop productivity.

RECEIVED November 2, 1976.

2

Chemical Enhancement of Sucrose Accumulation in Sugarcane

LOUIS G. NICKELL

Research Division, W. R. Grace & Co., Columbia, Md. 21044

*A decade of screening and evaluation in Hawaii has pro-
duced a surprising number of compounds with divergent
chemical structures which can increase the sugar content
of a cane crop at harvest. One compound is now registered
as a commercial product (Monsanto's Polaris). Two compounds are being evaluated under experimental permits
(American Cyanamid's Cycocel and Pennwalt's Ripenthol).
Several other products will move soon to the experimental
permit stage or be dropped. This group includes: Roundup
or one of its relatives, ethephon, asulam, MBR-12325,
Cetrimide, and Hyamine 1622. Another group with as
much potential but still in the early stages of testing includes: penicillin, bacitracin, n-valeric acid or one of its
relatives, vanillin, 6-azauracil, several of the furans, tetra-
hydrobezoic acid, and cacodylic acid.*

R ipening of the sugarcane plant is considered one of the most important aspects of sugar production from both a research and an opera-
tional point of view. Cane ripening is an extremely complex phenomenon.
Many endogenous factors in the growth and metabolism of the cane
plant are involved as well as a number of environmental factors. Alexander (1) has described ripening in sugarcane as fulfilling the potential
which has been created for "massive sugar accumulation in the storage
tissue previously laid down." Ripening is more commonly and simply
described as "maximizing sucrose and minimizing all other soluble solids
at harvest."

The use of chemicals to increase the content of sucrose at harvest is
not a new concept. Several references appeared decades ago in the
literature suggesting this approach; however, in most instances, the sug-

gestion was followed by the observation that too little was known about the physiology and biochemistry of the cane plant for such a program to be effective.

The potential value of being able to control the maturation of sugarcane at harvest is acknowledged throughout the sugarcane world. Environmental conditions make artificial control more valuable in some areas than in others. For example, the high quality of cane obtained in Queensland is ascribed largely to a change from normal, rapid growth of the crop to a prolonged period of slow growth induced by low temperatures (24) that override any effects of moisture availability (23). Most other areas are not so fortunate, having to contend with adverse interactions of sunlight, rain, temperature, salinity, and others (3, 4). Besides environmental variables, Hawaii has another problem as far as ripening is concerned since the year-round harvesting is dictated by the economics of the industry there.

Spasmodic reports on the use of plant growth regulators have appeared in the literature since the report of 2,4-D as the first material to be effective for this use. This pioneering work by Beauchamp in Cuba (2) in 1949 was followed by a series of random studies using materials available at that time, many of which were herbicides, enzyme inhibitors, metabolic inhibitors, chelating agents, and other types of chemicals with biological activity. No large-scale, serious program was launched, however, until basic studies in translocation concerned with the defoliation of sugarcane furnished the basis for initiating such a program (5, 6). Programs similar to that originating in Hawaii (7, 8) have been started at one time or another in Australia, South Africa, Puerto Rico, Florida, Louisiana, Taiwan, the Philippines, Mauritius, Rhodesia, and other sugar-producing countries.

Because of space this discussion is limited primarily to the results obtained in Hawaii. Particularly pertinent references to other work in other areas are included. Since the primary object of this presentation is to discuss positive field results, I describe briefly the screening method used to obtain the initial results. This test is very simple (7), consisting of adding test materials in solution or suspension by pipette or by needle and syringe into the whorl of leaves at the growing tip of the sugarcane stalk, which is field grown and near the right stage for normal maturation. Four weeks and five weeks after application of the test material, five to 10 stalks are harvested, analyzed, and compared with an untreated group of stalks. Large scale tests are applied by air. The effectiveness of the compound as a ripener is based on its ability to increase the quality of the treated stalks according to two major parameters for sugar production: (a) juice purity and (b) sugar as a percent of field cane weight. Juice purity is the percent of soluble solids in cane juice that is sucrose. The results are given by chemical groupings and relationships rather than in chronological order or on the basis of activity.

Phenoxy Compounds

The two most common, most widely used, and best known of the phenoxy compounds—2,4-dichlorophenoxyacetic acid (2,4-D) and 2-methyl-4-chlorophenoxyacetic acid (MCPA)—were studied in early investigations of the ripening of sugarcane with chemicals. The original report of positive activity by Beauchamp (2) was with 2,4-D. In our early studies with 2,4-D using the tests described above, there was slight activity although it varied from test to test. The use of an amine salt of 2,4-D gave more consistent results than did the acid. Compounds much more active than 2,4-D or its amine salts, however, were quickly found, and no advanced work was ever done with this material. MCPA was used primarily in combination with 2,3,6-trichlorobenzoic acid (TBA) in a formulation used in the topics known as Pesco 1815. This was active in tests carried out in Trinidad and other sugar-growing areas. I believe that the primary activity in this combination resided in the TBA, and, therefore, work with this material was continued (*see* next section) and that with MCPA dropped.

Benzoic Compounds

Trysben. Because of the early positive results mentioned above with Pesco 1815, Trysben (the dimethylamine salt of 2,3,6-trichlorobenzoic acid) was studied extensively in the early stages of the work in Hawaii.

2,3,6-Trichlorobenzoic Acid

This material was found to be exceptionally consistent in its activity and of sufficiently high activity to warrant serious consideration as a product candidate. However, a number of disadvantages precluded its use earlier as a commercial material (9). These disadvantages include a large chemical and biological stability. It is so large that the material can be taken into the plant, moved through the plant (as well as into the soil), and as a result of not being broken down, be available in the crop as a residue at a level greater than the probable maximum tolerance allowed on the basis of the toxicology of the compound. Also, Trysben is a hormone-type herbicide not unlike 2,4-D in some of its effects on many plants. This means that the restrictions involved using 2,4-D in Hawaii and other places would be the same for Trysben. Finally, there is no clear proprietary position among the companies concerned for the use of Trysben in ripening, and probably none of the manufacturers would have been interested in the expense necessary for its clearance. It has continued to be used routinely in our screening tests as a standard, however, because of its consistent activity and its availability as a formulated product. Because of this consistent, reasonably high activity, hundreds of other substituted benzoic acid compounds have been evaluated. Interestingly enough, only one or two have beeen close to Trysben in activity.

Mono-Substituted Benzoic Compounds. Of the numerous mono-substituted benzoic acids tested, four were found to have much higher activity than the rest. These have been selected for advanced field testing and include: 2-chlorobenzoic acid, 3-hydroxybenzoic acid, 3-cyanobenzoic acid, and 4-methoxybenzoic acid. (Although numerous di-substituted benzoic acid compounds have been screened, none has been outstanding.)

2-Chlorobenzoic Acid

Tri-Substituted Benzoic Compounds. In addition to Trysben, two of the more active tri-substituted materials are 3,6-dichloro-*o*-anisic acid (the herbicide Dicamba) and its methyl ester, disugran, sold under the trade name Racuza. Although Racuza advanced to the stage of extensive field testing in several areas throughout the world, including Hawaii, the inconsistent results obtained lessened the interest in this material.

Methyl–3,6-dichloro–*o*-anisate (disugran, Racuza)

Tetrahydrobenzoic Acid. This material, also known as 3-cyclohexene-1-carboxylic acid, has been evaluated in numerous screening tests and was found to have considerable activity in each. It has not advanced to the field testing stage at the present time.

Vanillin. Vanillin (4-hydroxy–3-methoxybenzaldehyde) is a well-known, naturally occurring compound present in small quantities in many plants, particularly the pod of the vanilla orchid. It is also found in potato parings, in sugar beets, in balsams, and in other natural oils and resins. Vanillin can be synthesized, and it is expected that field tests with this material will be started in the foreseeable future.

4- Hydroxy–3-methoxybenzaldehyde (vanillin)

Phthalic Compounds

Ripenthol. This material, also sold under the trade name Hydrothol-191 as an aquatic herbicide, is the monoamine salt of 7-oxabicyclo-(2,2,1)heptane–2,3-dicarboxylic acid (Endothal). Ripenthol was one of the first materials found to have significant activity in our ripening tests (7). This compound was advanced to microplot and large-plot tests, eventually going to larger scale and air applications. It has been registered as an experimental material for evaluation on more than 2000 acres of cane in Hawaii. Numerous relatives of this material were tested in the early screening stages. The results can be summarized as follows: the acid itself is active but in practical terms has very low activity; all inorganic salts show a negative activity; those metal salts tested have a fairly strong negative activity; all amine salts tested were more active than the acid, and the mono-substituted were more active than the di-substituted. Since Ripenthol has considerable phytotoxic activity, care must be taken in its administration. In other words, drift can be a particularly important problem for this material.

Mono–N,N-dimethyldodecylamine Salt of Endothal, Ripenthol

Quaternary Ammonium Compounds

N,N-Dimethyl–N-(2-hydroxyethyl)–N-octadecylammonium Chloride. This material was one of the first active compounds to proceed to the field testing stage (*10, 11*). Unfortunately, it did not show sufficient positive effects when applied by air to continue to be considered as a product candidate although it was active in all of the earlier stages of testing.

$$\left[\begin{array}{c} C_2H_4OH \\ | \\ CH_3-N-C_{18}H_{37} \\ | \\ CH_3 \end{array} \right]^+ Cl^-$$

N,N-Dimethyl–N-(2-hydroxyethyl)–N-octadecylammonium Chloride

2-Chloroethyltrimethylammonium Chloride (Cycocel, CCC). This well-known plant growth regulator, sold in Europe as chlormequat and in the United States and other parts of the world as Cycocel, is probably the most widely used plant growth regulator in the world (*12*). Its primary use is on wheat, causing a shortening of the stalk, resulting in less damage from rain, wind, and other inclement weather conditions, thus increasing the yield at harvest. It has survived all of the preliminary testing stages and is registered with the EPA as an experimental compound for evaluation as a ripener for sugarcane on more than 1000 acres in Hawaii.

$$\left[\begin{array}{c} CH_3 \\ | \\ CH_3-N-C_2H_4Cl \\ | \\ CH_3 \end{array} \right]^+ Cl^-$$

2-Chloroethyltrimethylammonium Chloride (chlormequat, Cycocel)

Hexadecyltrimethylammonium Bromide (Cetrimide). This quaternary compound is one of the most widely known of the "quats." It has been used for many years in medicine for diagnostic purposes and has other varied uses. Its activity as a ripener in sugarcane is significant. It is now at the most advanced field testing stage—i.e., airplane application to small blocks. It is of particular use in certain sugarcane growing areas such as Hawaii where current irrigation methods are being changed to trickle and sub-surface irrigation. The material is not inactivated by contact with soil as are most of its more serious competitors. It has been tested in at least two major field experiments and shown to be active when applied by sub-surface irrigation for the ripening of two different varieties under field conditions (*10, 11*).

$$\left[\begin{array}{c} CH_3 \\ | \\ CH_3\!-\!N\!-\!C_{16}H_{33} \\ | \\ CH_3 \end{array} \right]^+ Br^-$$

Hexadecyltrimethylammonium Bromide (Cetrimide)

Hyamine 1622 (Diisobutylphenoxyethoxyethyldimethylbenzylammonium Chloride). This material has shown activity of the same type and to the same degree as Cetrimide. It has not, however, been evaluated under field conditions as extensively.

$$\left[\begin{array}{c} (CH_3)_2CH_2CH_2 \\ \\ (CH_3)_2CH_2CH_2 \end{array} \!\!- \bigcirc \!\!- O\!-\!CH_2CH_2\!-\!O\!-\!CH_2CH_2\!-\!\overset{\displaystyle CH_3}{\underset{\displaystyle CH_3}{N}}\!-\!CH_2\!\!-\!\bigcirc \right]^+ Cl^-$$

Diisobutylphenoxyethoxyethyldimethylbenzylammonium Chloride

Organic Phosphorus Compounds

(2-Chloroethyl)phosphonic Acid. This plant growth regulator, known as ethephon and sold by two companies under the trade names, Ethrel and Cepha, is quite active as a ripener. Its activity was determined many years ago, but the price was thought to be too high for continued investigation at that time. Now, almost a decade later, having found it to be active for a number of other uses, there has been renewed interest in reevaluating it economically for its potential as a product candidate (*15*). Because of this hiatus in its testing program, this material is still in the screening stage in Hawaii.

$$ClCH_2CH_2\,P\!\!\overset{\displaystyle \overset{O}{\|}}{\underset{\displaystyle OH}{\diagup}}\!\!{\diagdown}\,OH$$

2-Chloroethylphosphonic Acid (ethephon)

N,N-Bis(phosphonomethyl)glycine. This substituted amino acid, originally given the code designation CP-41845 (*16*), and now given the generic name, glyphosine, is the well known Polaris, the first registered material for commercial use as a sugarcane ripening chemical. It has been evaluated as an experimental material on over 20,000 acres of sugarcane in Hawaii and has shown excellent results—about 10–15% yield increase—which is over one ton of sucrose per acre when applied to certain varieties grown on the rainy coasts of the island of Hawaii. More recent work has shown that varieties previously thought to be non-responsive to this ripener have been found to respond positively when surfactants are added to the formulation (*17*). Although it is not yet established how it relates to the mode of action of Polaris, the effect on slowing terminal growth is significant (*16*).

$$\text{HO}-\overset{\displaystyle O}{\overset{\|}{\text{C}}}-\text{CH}_2-\text{N}-\left[\text{CH}_2\overset{\displaystyle O}{\overset{\|}{\text{P}}}\underset{\text{OH}}{\overset{\text{OH}}{<}}\right]_2$$

N,*N*-Bis(phosphonomethyl)glycine (Polaris)

N-(Phosphonomethyl)glycine. This material, closely related chemically and probably biologically to Polaris, has been given the generic name of glyphosate. As the isopropylamine salt, it is the very potent herbicide Roundup which is especially active on grasses. It has been tested as the isopropylamine salt, as the acid, and in several other forms as a ripener on sugarcane and was found to be extremely active—in fact, much more active than Polaris itself.

$$\text{HO}-\overset{\displaystyle O}{\overset{\|}{\text{C}}}-\text{CH}_2-\text{NH}-\text{CH}_2\overset{\displaystyle O}{\overset{\|}{\text{P}}}\underset{\text{OH}}{\overset{\text{OH}}{<}}$$

N-(Phosphonomethyl)glycine

5-Chloro-2-thenyl-tri-*n*-butylphosphonium Chloride. Quite active in screening tests and in small-sized plot tests, this plant growth regulant (*13*) did not show sufficient activity in larger-scale tests to remain an active product candidate.

$$\text{Cl}-\boxed{}_{\text{S}}-\text{CH}_2\text{P}^+(\text{CH}_2\text{CH}_2\text{CH}_2\text{CH}_2)_3\text{Cl}^-$$

5-Chloro–2-thenyl–tri-*n*-butylphosphonium Chloride

Other Metal Organics

Arsenic. Dimethylarsenic acid (cacodylic acid), an organic arsenical herbicide, and its sodium salt were found to have considerable activity, both together and individually, in the ripening of sugarcane in preliminary screening tests. No advanced work has been done yet with these materials.

$$CH_3 \diagdown \quad \diagup O$$
$$As$$
$$CH_3 \diagup \quad \diagdown OH$$

Hydroxydimethylarsine Oxide (dimethylarsenic acid, cacodylic acid)

Surfactants

The use of surfactants as adjuvants in formulating pesticides for application to plants is well known and is a fairly advanced science. It was surprising, therefore, to find that when the amount of these materials was substantially increased, some of them, alone, had impressive ripening properties.

Tweens and Tergitols. Although the bulk of the testing has been done with Tween-20, all members of this series tested are active. Members of the old series of Tergitols, the non-biodegradable group, in particular NPX, were found to be active as ripeners when used alone. Also, the new biodegradable S series, as a group, were found to be active although certain members of this group were not active.

Aerosols and the Triton-X Series. Aerosol OT was evaluated as a representative of this group and was found to have slight activity—not sufficient to warrant additional testing. Members of the Triton-X series had no ripening activity.

Saturated Fatty Acids

To be sure that we considered this group of materials, several were evaluated for ripening activity. Surprisingly, those aliphatic monoacids having five or less carbon atoms showed considerable activity whereas those having six or more carbon atoms either showed insufficient or no activity. In addition, of course, these longer chain acids are practically insoluble in water. The peak of activity seems to be in the butyric, isobutyric, and valeric area. Although there is activity with formic acid, there would be difficulty in using it as a product since it is a volatile liquid, gives off disagreeable fumes, and produces superficial blisters on contact with the skin, being an active caustic. Diluted, it is locally an irritant and an astringent. Both formic and acetic acids are given severe animal toxicity ratings for acute local and for acute systemic. Propionic acid, although active in ripening, shows variable activity and inconsistent results; therefore, it probably is not a likely candidate for commercial use. The alkaline metal salts and certain esters, especially the ethyl esters, of these aliphatic monoacids with one to five carbon atoms also produce ripening activity. In many instances, the salts and esters do not possess the adverse properties, particularly the odors, that accompany the acids.

$$CH_3(CH_2)_3COOH$$

Valeric Acid (pentanoic acid)

Pyrimidines

6-Azauracil. Early in the testing program 6-azauracil and its riboside 6-azauridine showed considerable ripening activity. Cost estimates have prevented our pursuing these materials beyond the preliminary stages.

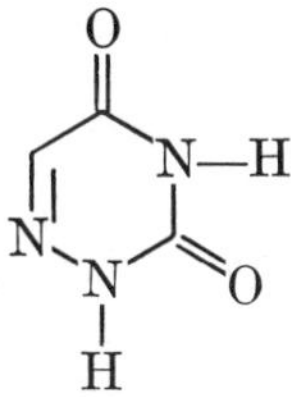

6-Azauracil

Laurylmercaptotetrahydropyrimidine. This material was one of the first compounds found active in the ripening screening program (*18*). It is the member of the group showing maximum activity for the control of plant rusts (*19*). Unfortunately, when it was taken to the field for advanced testing, the amount of activity shown did not hold up in commercial terms with a number of competitive compounds. Consequently, it has been dropped from the program.

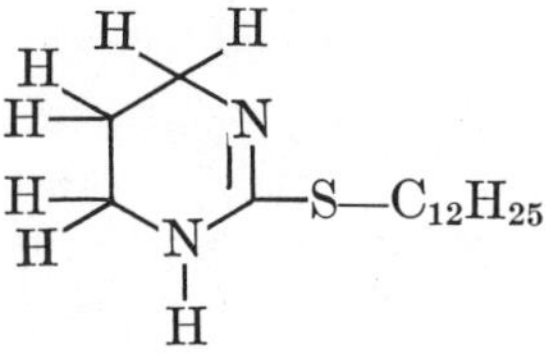

Laurylmercaptotetrahydropyrimidine

Other Substituted Pyrimidines. A number of substituted pyrimidines have been evaluated for their ripening effects and were found to be active. None has been followed up in the practical sense because of of cost estimates. The group includes:

> 6-Azacytosine
> 5-Nitrocytosine
> 4-Chloro-2,6-diaminopyrimidine
> 2-Amino-6-chloro-4-pyrimidinol
> 6-Amino-2-(ethylthio)-4-pyrimidinol
> 2-Amino-4-chloro-6-methylpyrimidine
> 6-Chloro-2,4-dimethyoxypyrimidine

Carbamates

Although numerous fungicides and herbicides fall into this class, it is interesting that only one carbamate has been found to have sufficient activity for advanced testing as a ripener of sugarcane (*15*). This material, methylsulfanilylcarbamate, has the generic name of asulam and is sold under the trade name of Asulox. It has shown activity in all the preliminary testing and is now at the stage of air application to determine whether it should be seriously considered as a commercial product candidate.

$$H_2N-\langle\rangle-SO_2NHCOOCH_3$$

Methylsulfanilylcarbamate (asulam, Asulox)

Cyclic Nitrogen Compounds

Pyridines. A pyridinol (2,3,5-trichloro–4-pyridinol) known as Daxtron and a pyridone (the sodium salt of 3-carboxyl-1-(*p*-chlorophenyl)-4,6-dimethyl-2-pyridone) known by the designation RH-531, both showed activity in screening tests. Neither of these pyridines has been carried to an advanced stage.

Sodium 3-Carboxyl–1-(*p*-chlorophenyl)–4,6-dimethyl–2-pyridonate

Pyridazines. 3-(2-Methylphenoxy)pyridazine, known as Credazine, was found to have sufficient activity in a screening test to warrant advanced tests. In this case again, the compound has considerable activity in all the preliminary tests but does not warrant continued interest after large-scale application.

3-(2-Methylphenoxy)pyridazine

Picolines. Both 2-picoline–N-oxide and Tordon (the potassium salt of 4-amino–3,5,6-trichloropicolinic acid) were found to have considerable activity in screening tests. Neither material has been given any advanced testing—Tordon for the same reason that prevented Trysben from becoming a serious product candidate (8).

4-Amino–3,5,6-trichloropicolinic Acid (picloram, Tordon)

Furans and Organic Amines

Many furans and tetrahydrofurans were found to have exciting activity as ripeners for sugarcane. Among these, especially good results were obtained with tetrahydrofuroic acid hydrazide.

Tetrahydrofuroic Acid Hydrazide

Two substituted toluidides were found to have good ripening activity. These are the grass inhibitor, Sustar (3'-(trifluoromethylsulfonamido)–p-acetotoluidide) and its close relative, designated MBR-12325. Sustar was tested originally, found to be active, and progressed through the various stages up through application to several multi-acre blocks. In the meantime it was found that MBR-12325 has considerably more activity than

Sustar. Preference is now being given to the latter compound in field testing.

3-(Trifluoromethylsulfonamido)–p-acetotoluidide (fluoridamid, Sustar)

Antibiotics and Derivatives

Isoaureomycin. Because of its known activity as a plant growth stimulator (20), isochlorotetracycline (isoaureomycin) was tested early in this program and was found to have considerable activity as a sugar-cane ripener. Unfortunately, its cost was too high to make it competitive as a commercial product.

Isochlorotetracycline (isoaureomycin)

Actidione and Anisomycin. These two anti-fungal antibiotics were tested, not because they were antibiotics, but because of their known inhibitory effects on biological systems such as those involved in protein synthesis. Both were found to be active (21) but were never pursued beyond the initial screening stages because of high cost and particularly their animal toxicity, especially to the mucosa.

Miscellaneous Antibiotics. Because of the activity found in some of the materials in the early testing and that are discussed later under Bacitracin and Penicillin, a considerable number of antibiotics were screened with the hope that there would be a range of activity which might relate (a) to their antimicrobial activity, (b) to their mode of action as antibiotics, or (c) to some other biological relationship. The results of some of this work are shown in Table I (22). Our hopes were substantiated in that we have found a range of activity going from

Table I. Antimicrobial Compounds Grouped According to Ripening Effects on Sugarcane (22)

High Activity

	Times Tested		*Times Tested*
Naramycin A	4	Streptomycin SO_4	2
Cycloserine	5	Rifamycin	2
Magnamycin	4	Novobiocin	5
Nystatin	1	Neomycin SO_4	4

Marginal Activity

Tylosin (base)	4	Gramicidin	5
Nisin	2	Terramycin	2

No Activity

Nalidixic acid	1	Sulfanilamide	2
Aureomycin	2	Polymyxin B-SO_4	1
Tetracycline	2	β-apo-Terramycin	1
Oleandomycin	1	Lincomycin	1
Isonicotinic hydrazide	1	Griseofulvin	1
Chloramphenicol	2	Hadacidin	1
Sulfadiazine	2	Tyrothricin	1
		Erythromycin	1

Hawaiian Planter's Record

negative to strongly positive. We intend to use this information to help us in our studies of mechanism of action.

Bacitracin. This polypeptide antibiotic and its zinc salt were both found to have considerable activity as sugarcane ripening agents. Efforts are now being made to obtain material in a form usable for agricultural purposes so that advanced block testing can be carried out.

Penicillin. Penicillin was tested for a number of reasons, not the least being that it is one of the few antibiotics that is available in bulk at a price which might be competitive with existing products and product candidates. Penicillins G and V, whether in their potassium or procain forms, were highly active as sugarcane ripeners. More surprisingly, 6-aminopenicillanic acid (6-APA) showed considerable activity itself, demonstrating that the antimicrobial activity of these materials is not

$$\langle\text{phenyl}\rangle\text{—CH}_2\text{CONH—CH—CH}\overset{\text{S}}{\diagdown}\text{C(CH}_3)_2$$
$$\overset{|}{\text{CO—N}}\overset{|}{\text{——CHCOOH}}$$

Penicillin G

related to their ripening activities. Furthermore, when the molecule is cleaved to even a greater degree, resulting in the formation of penicillamine, this material, having no biological activity in the microbial sense, is quite active as a sugarcane ripener. We feel that this latter information will have two major impacts upon this program. One is concerned with the chemical structure–biological activity relationships of compounds whose structures are known and whose antimicrobial activities have been studied in great detail. The other is that we have reduced the active molecules in this case to relatively simple chemistry. Now there is hope that a synthetic approach to related compounds which also might be active has potential.

Summary and Conclusions

The results of a little more than a decade of screening have shown a surprising number of chemical structures to have activity in increasing sugar accumulation in sugarcane. As in all such programs, many compounds active at the bench or greenhouse stage do not stand up under the requirements of field use. At present, one compound is registered as a commercial product (Monsanto's Polaris). Two compounds are being extensively evaluated under experimental permits (American Cyanamid's Cycocel and Pennwalt's Ripenthol). Several other products are in advanced testing stages and can be expected to move soon to the experimental permit stage or be dropped completely. This group now includes: Roundup or one of its relatives, ethephon, asulam, MBR-12325, Cetrimide, and Hyamine 1622. Another group with as much potential but not yet tested enough in the early stages for one reason or another includes: penicillin, bacitracin, n-valeric acid or one of its relatives, vanillin, 6-azauracil, several of the furans, tetrahydrobenzoic acid, and cacodylic acid. Of the major groups of herbicides, no triazine or substituted urea has yet shown sufficient activity to be of interest, even though many have been evaluated.

Several investigators from government, academe, and industry have expressed the view that the regulation of crop growth and metabolism may be the cause of one of the most important quantitative gains yet achieved in agriculture—a viewpoint with which I heartily agree. The success with sugarcane ripeners in giving yield increases over 10% is strong substantiation for such beliefs. The rapidly increasing industrial interest and action show that serious investigation into the potential of plant growth regulation is underway. If we are to succeed in the monumental task of producing the raw materials to satisfy the world's human energy requirements, we will need major spurts of achievement such as could be furnished by this approach to an array of crops. I expect increasing emphasis and support will be given to this important area of research because of its commercial potential.

Literature Cited

1. Alexander, A. G., "Sugarcane Physiology," Elsevier, Amsterdam, 1973.
2. Beauchamp, C. E., "A New Method of Increasing the Sugar Content of Sugarcane," *Proc. 23rd Ann. Mtg. Assoc. Tech. Azucareros, Cuba* (1949) 55–87.
3. Fogliata, F. A., Aso, P. J., "The Effects of Soil Soluble Salts on Sucrose Yield of Sugarcane," *Proc. 12th Cong. ISSCT*, Puerto Rico (1967) 682–694.
4. Fogliata, F. A., Dip, R. A., "Crecimiento y Maduracion de la Caña de Azucar en Tucuman," *Rev. Ind. Agric. Tucumann* (1967) **45** (3): 57.
5. Hartt, C. E., "Translocation of Sugar in the Cane Plant," *1963 Repts. Haw. Sugar Technol*, 151–167.
6. Hartt, C. E., Kortschak, H. P., Burr, G. O., "Effects of Defoliation, Deradication, and Darkening the Blade upon Translocation of C^{14} in Sugarcane," *Plant Physiol.* (1964) **39**, 15.
7. Tanimoto, T. T., Nickell, L. G., "Ripening Studies with Chemicals," *Ann. Rpt. Exp. Sta.*, Haw. Sugar Planters' Assoc. 1964: 1.
8. Nickell, L. G., Tanimoto, T. T., "Effects of Chemicals on Ripening of Sugarcane," *1965 Repts. Haw. Sugar Technol.*, 152–166.
9. Nickell, L. G., Tanimoto, T. T., "Sugarcane Ripening with Chemicals," *1967 Repts. Haw. Sugar Technol.* (1968) 104–109.
10. Tanimoto, T. T., Nickell, L. G., "Chemicals for Control of Ripening," *Ann. Rpt. Exp. Sta.*, Haw. Sugar Planters' Assoc. 1965: 1.
11. Nickell, L. G., Tanimoto, T. T., "Screening Finds New Ripeners—Most Active Undergo Field Tests," *Ann. Rpt., Exp. Sta.*, Haw. Sugar Planters' Assoc. 1967: 7.
12. Wittwer, S. H., "Growth Regulants in Agriculture," *Outlook Agric.* (1971) **6**, 205.
13. Nickell, L. G., "Chemical Ripeners for Sugarcane" (Proc. Subsurface and Drip. Irr. Seminar) *Univ. Haw. Coop. Ext. Serv. Misc. Publ.* (1973) **102**, 42.
14. Nickell, L. G., Takahashi, D. T., "Sugarcane Ripeners in Hawaii—1973," *1973 Rpts. Haw. Sugar Technol.* (1974) 76–84.
15. Nickell, L. G., Takahashi, D. T., "Field Studies with Sugarcane Ripeners in Hawaii—1974," *1974 Rpts. Haw. Sugar Technol.* (1975) 85–90.
16. Nickell, L. G., Takahashi, D. T., "Ripening Studies in Hawaii with CP-41845," *1971 Rpts. Haw. Sugar Technol.* (1972) 73–82.
17. Nickell, L. G., "Plant Growth Regulants in Sugarcane," *Bull. Plant Growth Regul.* (1974) **2**, 51.
18. Taminoto, T. T., Nickell, L. G., "Ripening with Chemicals," *Ann. Rpt., Exp. Sta.*, Haw. Sugar Planters' Assoc. 1966: 2.
19. Nickell, L. G., Goenaga, A., Gordon, P. N., "2-n-Alkylmercapto-1,4,5,6-tetrahydropyrimidines, Chemotherapeutic Agents for Plant Rusts," *Plant Dis. Rptr.* (1961) **45**, 756.
20. Nickell, L. G., "Plant Growth Stimulation," U. S. Patent 2,907,650, Oct. 6, 1959.
21. Nickell, L. G., Maretzki, A., "Sugarcane Ripening Compounds—Comparison of Chemical, Biochemical, and Biological Properties," *Hawaii. Plant. Rec.* (1970) **58**, 71.
22. Nickell, L. G., Takahashi, D. T., "The Effects of Antibiotics and Other Antimicrobial Agents on the Ripening of Sugarcane," *Hawaii. Plant. Rec.* (1975) **59**, 15.

23. Yates, R. A., "Studies on the Irrigation of Sugarcane," *Aust. J. Agric. Res.* (1967) **18**, 903.
24. Yates, R. A., " A Review of Some Recent Work on Chemical Ripening of Sugarcane," *Int. Sugar J.* (1972) **74**, 198.

RECEIVED September 22, 1976.

Fruit Abscission and Chemical Aids to Harvest

R. H. BIGGS and S. K. MURPHY

University of Florida, Gainesville, Fla. 32611

Chemical aids to harvest and fruit abscission agents are reviewed and discussed in relation to structure, mode of action, and predictability of response. Also, an attempt is made to assess the physiological basis for different responses among fruits of differing physiological age and among various organs on the same plant.

During the past two decades intensive research has been conducted on the phenomenon of organ abscission in plants. A basic knowledge of the processes gives insight into the metabolism of cell walls which has applicability in attempts to control the ripening process in many fruit production programs.

Prevention or induction of organ abscission is a primary advantage in regulating abscission of plant organs. Selecting the time and number of organs to be retained or abscised can influence the size and quality of a particular fruit crop. With most fruit, there is a definite need for controlling the number of flowers and young fruit which set early in the season (1) and for promoting abscission of mature fruits at desired harvest times later in the season (2). Thus one would like to have a chemical that would reduce the bonding force of fruit to stem, is predictable in its action, is relatively non-toxic to the consumer, and is not phytotoxic to the plant. If one is interested in the fresh fruit market, two other conditions are necessary: the fruit must be free of injury and, more desirably, colored.

Attention in this review is directed toward abscission of mature fruits. Since plants abscise organs naturally, we examine the nature of the abscission process from the standpoint of structural changes and enzymes involved; how plants seemingly regulate a selective control over abscission; natural and synthetic chemicals affecting abscission; whether a chemotherapeutic approach to fruit loosening is feasible; and, lastly, the basis for obtaining selective organ abscission.

The majority of data on the physiology of abscission were obtained using leaves, and principal concepts evolved from these observations. Differences among fruit, leaf, or other organ abscission are indicated as applicable to the discussion. Because of the vast number of articles published on abscission, neither time nor space will allow this to be an all-inclusive review, but hopefully the selected references allow an assessment of present knowledge of fruit abscission and chemical aids to harvest. Several excellent reviews are available on various phases of abscission (2–6) and allied areas (7, 8, 9).

Nature of Abscission Zones

Anatomy of the abscission zone at the base of mature fruits has been investigated in apples (10), cherries (11), citrus (12), and olives (13) and has been discussed as related to mechanical fruit removal (2, 14). The histochemistry of the separation layer of mature fruit is very similar to that of leaves of citrus (15) and bean (16, 17). The actual separation occurs through one or more processes in the separation zone: (a) a weakening of the cementing ability of the middle lamella between cells, and (b) a softening of the entire cells (17). In some plants, the abscission zone is structurally differentiated as a layer of compact cells or as a zone of localized cell division; in other species, abscission may occur across a transect of cells which show little or no visible differentiation (18). Thus, major components of the separation process involve the dissolution of middle lamella and lysis of cell walls or entire cells in the separation zone, but this is not necessarily associated with distinctive morphological characteristics (19).

Unique features of fruit abscission need emphasis here. The abscission zone at the base of citrus fruit pedicels in the early part of stage III of fruit development [see Bain (20) for a discussion of citrus developmental stages] is not the structurally weak point when the fruit is removed by a shearing force or a straight pull. A shearing force often leads to stemming (a term denoting pieces of stem still attached to fruit), and force applied longitudinally to the stem leads to plugging [a term denoting pieces of pericarp (rind) removed from the fruit]. This indicates that before abscission is induced, the pericarp or stem may be the weakest point even though there is less vascular tissue in the zone of separation. Cell wall changes are needed to weaken the bonding force of the tissue (1).

Secondly, in many fruits there are two abscission zones, one between the ovary and pedicel and the other at the pedicel–peduncle (rachis) junction. Citrus fruit is formed from a hypogynous pistil. Abscission of mature citrus fruit takes place at the pedicel–fruit junction (button area) and at the base of the pedicel in young fruits. The switch in zones of separation occurs in early stage I of fruit growth so separation subsequent to this developmental stage is primarily at the base of the fruit. In mature fruit the capacity for separation at the base of the pedicel no longer exists with most commercial cultivars of citrus (1).

Thirdly, the initial reduction in bonding strength in the separation zone at the base of the fruit is apparently associated with a swelling reaction of cell wall material. Similarly, anatomical and histochemical investigations of citrus leaf abscission by Hodgson (15) over 50 years ago pointed to this phenomenon. From free hand sections of living material, marked swelling and gelatinization of cell wall material prior to cell wall dissolution was observed.

Our present knowledge of the nature of chemical alteration of polymers indicates that a decrease in crosslinking between polymers and endobreaking of polymers results in swelling—hence, the attention to enzymes that catalyze these types of reactions in cell wall polymers.

Nature of the Reduction in Bonding Force

Cell wall dissolution in the separation layer of abscissing organs has been convincingly demonstrated (16). Two enzyme complexes have been implicated in this dissolution of cell walls during abscission of bean leaves, namely, those that act on polygalacturonic acid and on cellulose polymers. Morre (21), using cell-sloughing from cucumber pericarp explants as a test for pectinases, found an increase in pectinases of abscission zones of bean leaves as break-strength decreased. Inhibition of abscission with cycloheximide resulted in reduced pectinase activity. Other investigators (22, 23, 24, 25), using synthetic substrates of carboxymethylcellulose (CMC) and sodium polypectate to test for cellulase and pectinases, respectively, have demonstrated an increase in cellulases as the tensil strength of separation zones decreased while there was little change in pectinases.

Other enzymes known to degrade cell walls of higher plants do not change during abscission (22). Ethylene has been found to stimulate the formation or activity of a large number of enzymes including cellulases (7, 26) which have been shown to be strongly correlated with abscission processes (22, 23, 27). Ethylene affects bean leaf abscission zones by increasing cellulases and decreasing tensile strength (22). Similar effects have been demonstrated with abscisic acid (28). RNA (29) and protein synthesis accompany increases in cellulase activity. Inhibitors of protein synthesis, hence cellulases, inhibited abscission (22). These observations indicated *de novo* synthesis of cellulase concurrent with abscission, and, indeed, experiments designed to test for *de novo* synthesis of cellulase were positive (27). Tests with citrus indicate the same systems may be present in fruit abscission zones. Pollard and Biggs (24) demonstrated that an increase in cellulase activity is associated with natural abscission and with ethylene-stimulated abscission of citrus fruit.

Ethylene also increases polygalacturonase activity in abscission zones, and this may also contribute to the abscission of some organs (31, 32). The polygalacturonases isolated from plants, however, have been the exo-types (31), and there have been some questions concerning separation of cellulases from polygalacturonase on the basis of a salting procedure (33).

Adding to the complexity of cell wall changes during abscission is the observation that there are two forms of *endo*-cellulase (34). Apparently, however, only the extracellular form with a PI of 9.5 is associated with abscission (35). The form with a PI of 4.5 has been shown to be associated with the plasma membrane and may be changed to the 9.5-PI form upon secretion (36).

Chemical Stimulation of Abscission

A productive field of investigation to control abscission has been the search for natural and synthetic chemicals to prevent or hasten abscission. Reports of the many chemicals that alter the abscission processes are too numerous to mention. Some of those known to modify fruit abscission include: abscisic acid (4); auxins (2, 37); auxin antagonists (38); ethylene (7); ethylene-releasing chemicals (14); gibberellins (39); cytokinins (40); growth retardants (14); inhibitors of respiration (38), particularly sulfhydryl reactants (38); protein inhibitors (24, 41, 42); nucleic acid modifying agents (24, 41); minerals in near phytotoxic quantities (43, 44); ascorbic acid (42, 45, 46, 47, 48); and others that do not seem to fit any of the categories mentioned (38, 42). A decreasing number of chemicals accelerated fruit abscission as explants—branch- and whole-tree tests, respectively, were used (42).

The natural growth regulators that seem to play a role in abscission are abscisic acid (14, 28, 46, 49, 50, 51); auxin (2, 6, 37, 39, 46, 52–56, 109); cytokinins (2, 40, 57, 58); ethylene (7, 59); gibberellins (2, 39); and unidentified senescence and abscission accelerating agents (60, 61, 62, 63).

Out of the array of growth regulators that influence fruit abscission, ones that prevent abscission are fewer in number and seem to fall in the class of auxins (1, 64), gibberellins, or cytokinins (65). Auxin has both indirect (7, 66) and direct (6) effects on abscission, whereas cytokinins seem to act indirectly through modifying conditions in contiguous tissues (*40, 57*). Both auxin and cytokinins are components needed for healthy, growing tissues (*6*).

Auxins (*67*), gibberellins (*68*), cytokinins (*69*), and abscisic acid (*67*) can enhance the production of ethylene if added in concentrations that are generally considered stronger than tissue levels. Paradoxically, ethylene was one of the first chemicals identified as a potent defoliant (*70*), and now it has been shown to be a natural product of plant tissues that seems to regulate abscission (*68, 71*) as well as an influence on a host of other physiological reactions (*7*).

Ethylene is produced in measurable amounts in a number of fruits, leaves, and shoots under normal conditions (*7, 46, 59, 72*) and in large amounts after treatment with certain chemicals (*14, 43, 44, 46*), mechanical stresses (*72*), and adverse environmental factors (*73*). Thus, acceleration of abscission by many agents seems to occur via ethylene production—a fact that is being used to assay chemicals as potential accelerating agents for abscission (*74*). Apparently, this is the basis for the induction of abscission by placement of abscission chemicals on the surface of an organ such as an orange (*41*).

Ethylene-Generating Compounds

Several compounds have a pronounced effect on physiological mechanisms through a release of ethylene by chemical change of the compound applied. Ethylene chlorohydrin was one of the first compounds in this class to be researched for agricultural applications, and, even in the early work, it was noted that the active principle was ethylene (*7*). In the past several years, with the investigations on the action of ethylene on physiological mechanisms, including abscission, renewed interest in the ethylene-generating compounds, (2-chloroethane)phosphonic acid, ethephon (*75*), 2-chloroethyltris(2-methoxyethoxy)silane (*76*), and glyoxime (*77*) have been investigated for agricultural applications.

Ethephon has received the most attention for use in loosening fruit as an aid to mechanical harvesting. Various degrees of success have been obtained using ethephon on olives (*78, 79, 80*), blueberries (*81*), cherries (*82–86*), plums (*87*), peaches (*88, 89*), and apples (*90, 91, 92, 93*). The initial visible fruit response is an increase in coloration, but higher concentrations do result in abscission of the fruit.

Separately, and in combination with ethephon, succinic acid, 2,2′-dimethylhydrazide (SADH), has been tested as an effective coloring, and sometimes abscission, agent in cherries (*84, 94*), peaches (*89, 95–99*), apricots (*99*), and apples (*93, 100, 101, 102*). Combinations such as

2,4,5-TP, 2,4,5-T, or SADH with ethephon have been found to reduce excessive abscission caused by the latter.

General and specific problems on fruits are encountered in the use of chemicals to loosen fruit for harvest. Overall tree or branch growth and subsequent yields can be reduced by retardant-type chemicals. Yellowing and abscission of leaves often accompanies whole-tree spray treatments. Post-harvest fruit quality, however, is the most serious problem. Although good coloration is usually obtained, excessive and rapid fruit drop may occur, leading to accelerated softening or decay; hence, reduced storage life of chemically harvested fruit is common.

Ethylene-Injury Abscission Accelerating Agents

Cycloheximide (CHI), when applied directly to separation zones, will inhibit abscission of citrus fruit explants (24) as it does bean petiole explants (29). By application to citrus fruit surfaces, the rind is injured, evidenced as small pitted areas, and ethylene is produced in quantities that accelerate the abscission processes (41). Apparently, CHI applied to the fruit wall enhances the senescence processes, including ethylene production, thereby stimulating citrus fruit abscission via ethylene production and not by action directly on the tissues in the separation zone (41). Cycloheximide has also been used for harvest of olives (80, 103) and apples (104).

5-Chloro-3-methyl-4-nitro-1H-pyrazole (Release) has no auxin-, gibberellin-, or cytokinin-like activity, yet it is an effective abscission agent. At the present, there is no evidence to indicate that Release retards auxin, gibberellin, or cytokinin activity. It does stimulate and enhance the tissue production of ethylene (105). Release is fairly stable, and there is no indication that it is degraded by the tissue to ethylene per se (106).

A model for the mode of action is that it is fairly well absorbed through the cuticle and epidermal tissue of the peel of an orange. As shown by Murphy and Biggs (105) most of peel-applied Release remains in the flavedo with little in the albedo (white spongy portion of the peel). Much of its radioactivity can be shown to be present in the cold-pressed oil from the lysigenous glands. Data from using ^{14}C-labelled Release would seem to indicate that the glands become sites for the compound to accumulate apparently to a level where cells lining the gland wall are stimulated to produce ethylene at levels greater than would be induced by a mere stress reaction. Too low or too high a treatment concentration of Release results in reduced uptake into the flavedo although excessive amounts result in effective fruit abscission. This suggests some tissue damage may be required for active accumulation of Release.

Release must be present on the surface of the peel for several hours for absorption of threshold quantities to induce ethylene production to a level that initiates abscission. This accumulation depends on external factors, primarily humidity and temperature (*107*).

A recent development in the field of chemical abscission formulations has been the use of mixtures of chemicals to induce and to control fruit loosening. CHI and Release have been shown to have a synergistic influence on fruit abscission (*108*).

Conclusions and Prospectus

A major factor leading to abscission is the weakening of the middle lamella, cell walls, or cells in a separation zone across the petiole, pedicel, or stem. Although any of the known plant hormones can alter the progress of abscission, ethylene remains unique as the principal stimulus of the increased activity of wall-degrading enzymes in abscission, whereas auxin can be given a central role in the retardation of abscission. With the present level of understanding, it would seem that abscission control involves interactions between auxin and ethylene.

The chemotherapeutic approach to fruit lossening has succeeded with chemicals that stimulate target organs to produce ethylene or with organs that release ethylene as a degradation product inside the target organ. The next generation of chemical formulations will probably also induce ethylene production plus have a direct influence on abscission-zone cellulases or later auxin transport, production, activity, or degradation so that the quality of ethylene produced will be more effective. The possibilities in formulation of preferential wetting, uptake, and degradation of active principal between target and non-target organs should not be neglected.

Literature Cited

1. Biggs, R. H., *HortScience* (1971) **6**, 388–392.
2. Cooper, W. C., Rassmussen, G. K., Roger, B. J., Reece, P. C., Henry, W. H., *Plant Physiol.* (1968) **43**, 1560–1576.
3. Addicott, F. T., Lyon, J. L., "Shedding of Plant Parts," Kozlowski, T. T., Ed., 85–124, Academic Press, N.Y., 1973.
4. Carns, H. R., *Ann. Rev. Plant Physiol.* (1966) **17**, 295–314.
5. "Shedding of Plant Parts," Kozlowski, T. T., Ed., Academic Press, N.Y., 1973.
6. Rubinstein, B., Leopold, A. C., *Q. Rev. Biol.* (1964) **39**, 356–372.
7. Abeles, F. B., "Ethylene in Plant Biology," Academic Press, N. Y., 1973.
8. Addicott, F. T., Lyon, J. L., *Ann. Rev. Plant Physiol.* (1969) **20**, 139–164.
9. Weaver, R. J., "Plant Growth Substances in Agriculture," W. H. Freeman and Co., San Francisco, 1972.
10. Edgerton, L. J., *HortScience* (1971) **6**, 378–382.
11. Bukovac, M. J., *HortScience* (1971) **6**, 385–388.

12. Wilson, W. C., Hendershott, C. H., *Proc. Am. Soc. Hortic. Sci.* (1968) **92**, 203–210.
13. Reed, R., Hartmann, H. T., *J. Am. Soc. Hortic. Sci.* (1976) **101**, 724–730.
14. Cooper, W. C., Henry, W. H., "Shedding of Plant Parts," Kozlowski, T. T., Ed., pp. 475–524, Academic Press, New York, 1973.
15. Hodgson, R. W., *Univ. Calif. Berkeley Publ. Bot.* (1918) **6**, 417–428.
16. Webster, Barbara D., *Plant Physiol.* (1968) **43**, 1512–1544.
17. Webster, Barbara D., "Shedding of Plant Parts," Kozlowski, T. T., Ed., 45–83, Academic Press, N.Y., 1973.
18. Gawadi, A. G., Avery, G. S., *Am. J. Bot.* (1950) **37**, 172–180.
19. Leopold, A. C., *HortScience* (1971) **6**, 24–26.
20. Bain, J. M., *Aust. J. Bot.* (1957) **6**, 1–24.
21. Morre, D. J., *Plant Physiol.* (1968) **43**, 1545–1559.
22. Abeles, F. B., *Plant Physiol.* (1969) **44**, 447–452.
23. Horton, R. F., Osborne, D. J., *Nature (London)* (1967) **214**, 1086–1088.
24. Pollard, J. E., Biggs, R. H., *J. Am. Soc. Hortic. Sci.* (1970) **95**, 667–673.
25. Ratner, A., Goren, R., Monselise, S. P., *Plant Physiol.* (1969) **44**, 1717–1723.
26. Abeles, F. B., Forrence, L. E., *Plant Physiol.* (1970) **45**, 395–400.
27. Lewis, L. N., Varner, J. E., *Plant Physiol.* (1970) **46**, 194–199.
28. Craker, L. E., Abeles, F. B., *Plant Physiol.* (1969) **44**, 1144–1149.
29. Abeles, F. B., Holm, R. E., *Plant Physiol.* (1966) **41**, 1337–1342.
30. Abeles, F. B., Holm, R. E., *Ann. N.Y. Acad. Sci.* (1967) **144**, 367–373.
31. Riov, J., *Plant Physiol.* (1974) **53**, 312–316.
32. Greenberg, J., Goren, R., Riov, J., *Physiol. Plant.* (1975) **34**, 1–7.
33. Murphy, S. K., Biggs, R. H., *J. Am. Soc. Hortic. Sci.* (1977) in press.
34. Reid, P. D., Strong, H. G., Lew, F., Lewis, L. N. (1974) **53**, 732–737.
35. Lewis, L. N., Linkins, A. E., O'Sullivan, S., Reid, P. D., "Plant Growth Substances," 708–718, Hirokawa Publishing Co., Inc., Tokyo, 1973.
36. Koehler, D. E., Leonard, R. T., Vanderwoude, W. J., Linkins, A. E., Lewis, L. N., *Plant Physiol.* (1976) **58**, 324–330.
37. Laibach, F., *Ber. Dtsch. Bot. Ges.* (1933) **51**, 386–392.
38. Wilson, W. C., Hendershott, C. H., *Proc. Am. Soc. Hortic. Sci.* (1967) **90**, 123–129.
39. Lewis, L. N., Bakshi, J. C., *Plant Physiol.* (1968) **43**, 351–358.
40. Osborne, D. J., Moss, S. E., *Nature (London)* (1963) **200**, 1299–1301.
41. Cooper, W. C., Rassmussen, G. K., Hutchison, D. J., *BioScience* (1969) **19**, 443–444.
42. Wilson, W. C., *Proc. Fla. State Hortic. Soc.* (1969) **82**, 75–81.
43. Barmore, C. R., Biggs, R. H., *J. Am. Soc. Hortic. Sci.* (1970) **95**, 211–213.
44. Ben-Yehoshua, S., Biggs, R. H., *Plant Physiol.* (1970) **45**, 604–607.
45. Cooper, W. C., Henry, W. H., *Citrus Ind.* (1967) **48**, 5–7.
46. Palmer, R., Hield, H. Z., Lewis, L. N., *Proc. Int. Citrus Symp. 1st* (1969) **3**, 1135–1143.
47. Rasmussen, G. K., Jones, J. W., *Citrus Ind.* (1969) **50**, 26–28.
48. Wilson, W. C., Coppock, G. E., *Proc. Int. Citrus Symp. 1st* (1969) **3**, 1125–1134.
49. Böttger, M., *Planta* (1970) **93**, 205–213.
50. Cooper, W. C., Horanic, G., *Plant Physiol.* (1973) **51**, 1002–1004.
51. Liu, W. C., Carns, H. R., *Science* (1961) **134**, 384–385.
52. Addicott, F. T., Lynch, R. S., *Science* (1951) **114**, 688–689.
53. Biggs, R. H., Leopold, A. C., *Am. J. Bot.* (1958) **45**, 547–551.
54. Ismail, M. A., *J. Am. Soc. Hortic. Sci.* (1970) **95**, 319–322.
55. Jacobs, W. P., *Proc. Int. Hortic. Congr. 16th* (1962) 619–625.
56. La Rue, C. D., *Proc. Natl. Acad. Sci. U.S.A.* (1936) **22**, 255–259.
57. Scott, P. C., Leopold, A. C., *Plant Physiol.* (1966) **41**, 826–830.
58. Van Staden, J., *J. Exp. Bot.* (1973) **24**, 667–673.

59. Burg, S. P., *Plant Physiol.* (1968) **43**, 1503–1511.
60. Chang, Y., Jacobs, W. P., *Am. J. Bot.* (1973) **60**, 10–16.
61. Hall, W. C., Herrero, F. A., Katterman, F. R. H., *Bot. Gaz* (1961).
62. Osborne, D. J., Jackson, M., Milborrow, B. V., *Nature (London), New Biol.* (1972) **240**, 98–101.
63. Smith, O. E., *New Phytol.* (1968) **68**, 313–332.
64. Jacobs, W. P., *Plant Physiol.* (1968) **43**, 1480–1495.
65. Chatterjee, S. K., Leopold, A. C., *Plant Physiol.* (1964) **39**, 334–337.
66. Halloway, M., Osborne, D. J., *Science* (1969) **123**, 1067–1068.
67. Abeles, F. B., *Physiol. Plant.* (1967) **20**, 442–454.
68. Abeles, F. B., Rubinstein, B., *Plant Physiol.* (1964) **39**, 963–969.
69. Abeles, F. B., Holm, R. E., Gahagan, H. E., *Plant Physiol.* (1967) **42**, 1351–1356.
70. Neliubov, D., *Beih. Bot. Centralbl.* (1901) **10**, 128–139.
71. Hall, W. C., *Bot. Gaz.* (1952) **113**, 310–322.
72. Vines, H. M., Grierson, W., Edwards, G. J., *Proc. Amer. Soc. Hortic. Sci.* (1968) **92**, 227–234.
73. Addicott, F. T., *Plant Physiol.* (1968) **43**, 1471–1479.
74. Cooper, W. C., Henry, W. H., Hearn, C. J., *Proc. Fla. State Hortic.* (1970) **83**, 89–92.
75. de Wilde, R. C., *HortScience* (1971) **6**, 364–367.
76. Hartman, H. T., Reed, W., Opitz, K., *J. Am. Soc. Hortic. Sci.* (1976) **101**, 278–280.
77. Wilcox, M., Taylor, J. B., Wilson, W. C., Chen, I. Y., *Proc. Fla. State Hortic. Soc.* (1974) **87**, 22–24.
78. Vitagliano, C., *J. Am. Soc. Hortic. Sci.* (1975) **100**, 482–484.
79. Ibid., **101**, 591.
80. Hartman, H. T., Tombesi, A., Whisler, J., *J. Am. Soc. Hortic. Sci.* (1970) **95**, 35–40.
81. Ismail, M. A., *HortScience* (1974) **9**, 3.
82. Schumacher, R., Frankhauser, F., *Schwiez. Z. Obst. Weinbau.* (1969) **105**, 596–605.
83. Bukovac, M. J., Zucconi, E., Larsen, R. P., Kesner, C. D., *J. Am. Soc. Hortic. Sci.* (1969) **94**, 226–230.
84. Looney, N. E., McMechan, A. D., *J. Am. Soc. Hortic. Sci.* (1970) **95**, 252–255.
85. Aebig, D. E., Dewey, D. H., *HortScience* (1974) **9**, 5.
86. Wittenbach, V. A., Bokovac, M. J., *J. Am. Soc. Hortic. Sci.* (1975) **100**, 302.
87. Bukovac, M. J., *Ann. Rep. Mich. State Hortic. Soc.* (1969) **99**, 32–34.
88. Buchanan, D. W., Biggs, R. H., *J. Am. Soc. Hortic. Sci.* (1969) **94**, 327–329.
89. Looney, N. E., *Can. J. Plant Sci.* (1972) **52**, 73–80.
90. Edgerton, L. J., *Proc. N.Y. State Hortic. Soc.* (1968) **113**, 99–102.
91. Edgerton, L. J., Greenhalgh, W. J., *J. Am. Soc. Hortic. Sci.* (1969) **94**, 1–13.
92. Edgerton, L. J., Hatch, A. H., *Proc. N.Y. State Hortic. Soc.* (1969) **114**, 109–113.
93. Lord, W. J., Greene, D. W., Damon, Jr., R. A., *J. Am. Soc. Hortic. Sci.* (1975) **100**, 259.
94. Chaplin, L. T., Kenworthy, A. L., *J. Am. Soc. Hortic. Sci.* (1970) **95**, 532–536.
95. Beyers, R. E., Emerson, F. H., *J. Am. Soc. Hortic. Sci.* (1969) **94**, 641–645.
96. Beyers, R. E., Emerson, F. H., Dostal, H. C., *J. Am. Soc. Hortic. Sci.* (1972) **97**, 420.

97. Gambrell, C. E., Sirus, E. T., Stembridge, G. E., Rhodes, W. H., *J. Am. Soc. Hortic. Sci.* (1972) **97**, 268–272.
98. Baumgardner, R. A., Stembridge, G. E., Van Blaricon, L. O., Gambrell, C. E., *J. Am. Soc. Hortic. Sci.* (1972) **97**, 485–488.
99. Guelfat-Reich, S., Ben-Arie, R., *J. Am. Soc. Hortic. Sci.* (1975) **100**, 517–519.
100. Edgerton, L. J., Hoffman, M. B., *Proc. Am. Soc. Hortic. Sci.* (1965).
101. Edgerton, L. J., Blanpied, G. D., *J. Am. Soc. Hortic. Sci.* (1970) **95**, 664–667.
102. Looney, N. E., *J. Am. Soc. Hortic. Sci.* (1971) **96**, 350.
103. Hartman, H. T., El-Hamady, E., Whisler, J., *J. Am. Soc. Hortic. Sci.* (1972) **97**, 781–785.
104. Edgerton, L. J., *Proc. N.Y. State Hortic. Soc.* (1970) **115**, 163–166.
105. Murphy, S. K., Biggs, R. H., *Proc. Fla. State Hortic. Soc.* (1976) in press.
106. Winkler, V. W., Wilson, W. C., Kenney, D. S., Yoder, J. M., *Proc. Fla. State Hortic. Soc.* (1974) **87**, 321–324.
107. Murphy, S. K., Winkler, V. W., Biggs, R. H., *HortScience* (1976) **11**, 32.
108. Davies, F. S., Cooper, W. C., Galena, F. E., *Proc. Fla. State Hortic. Soc.* (1975) **88**, 107–113.
109. Rubinstein, B., Leopold, A. C., *Plant Physiol.* (1963) **38**, 262–267.

RECEIVED February 9, 1977. Florida Agricultural Experiment Stations' journal series No. 435.

Modification of Growth Regulatory Action with Inorganic Solutes

A. CARL LEOPOLD

University of Nebraska, Lincoln, Nebr. 68583

Actions of each of the plant growth hormones can be altered by calcium salts; action of hormones may be either promoted or inhibited by calcium. In some instances ammonium salts can have effects opposite to those of calcium, suggesting involvement of salts on membranes and macromolecules in the Hofmeister series. The idea that membranes are altered by salts is extended by evidence that $CaCl_2$ can alter the specific binding of auxin to membrane pellets from corn coleoptiles. It is proposed that the actions of natural hormones and of exogenous growth regulators may be subject to regulation by solutes which contribute to the characteristics of plant membranes.

In regulatory biology there are striking differences in the responsiveness of different tissues, organs, or species to hormones or to synthetic regulators. Until now little attention has been given to the question of why such strong differences in responsiveness exist or even to the question of why individual tissues go through sequences of changing responsiveness to different hormones. It is presumed that the plant hormones act as regulators through their attachment to some sites of action and that these sites may be located on membrane surfaces such as the plasmalemma (1). This review calls attention to the possibility that responsiveness of plants to regulators, either natural hormones or synthetic regulators, may be altered by chemical species that change the configuration of membranes or macromolecules in such a way that the attachment of the hormones or regulators to sites of action on the membranes or macromolecules may be altered.

The best-known solute effects on membrane and macromolecule properties are those of the inorganic salts. Hofmeister (2) showed that some inorganic salts had strong solubilizing effects on proteins whereas other salts had strong desolubilizing effects; the range of effects of various salts is known as the "Hofmeister series," with strong solubilizing effects at one end of the series (e.g., $CaCl_2$) and strong desolubilizing effects on the other end (e.g., $(NH_4)_2SO_4$). The series has been described in more contemporary terms as ranging between destabilizers and stabilizers (3). The effects are a consequence of alterations of charged groups on the surface of the macromolecules and on lyotropic effects or alterations of the structural interactions of water molecules with the macromolecule and with one another in the water lattices that form around the macromolecules. Stabilization and destabilization effects may alter the properties of nucleic acids (e.g., changing melting curves), may alter the activities of enzymes not only with respect to the extent of enzymatic activity but even the substrate specificity, and may alter the permeability of membranes (3).

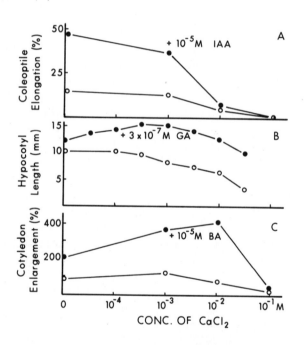

Plant Growth Substances, 1973

Figure 1.   Effects of $CaCl_2$ on bioassays for auxin, gibberellin, and cytokinin. A: effects of $CaCl_2$ on elongation of oat coleoptile sections in the presence and absence of indoleactic acid; B: effects on elongation of lettuce hypocotyls in the presence and absence of gibberellic acid; and C: effects on enlargement of Xanthium cotyledon pieces in the presence and absence of benzyladenine (13).

There have been frequent reports that inorganic salts can alter regulatory events in plants, but the reports have received relatively little attention. For example, calcium inhibits the auxin stimulations of growth (4), and in fact, Thimann and Schneider (5) suggested that the effects of various salts on auxin-stimulated growth might be related to their position in the Hofmeister series. Again, calcium stimulates root growth (6). More recently, calcium has been shown to alter the hormonal controls of leaf senescence (7) and of abscission (8). Several ethylene effects have been shown to be depressed by calcium salts (8, 9, 10). Even flowering has been reported to be altered by inorganic salts—calcium promotes (11) and ammonium inhibits (12). This abbreviated review shows that inorganic salts may alter numerous hormonal and developmental functions in plants, that calcium has a special degree of effectiveness, and that ammonium may have effects opposite to those of calcium.

Leopold et al. (13) examined the effects of calcium salts on each of the known plant hormones. They showed that in addition to the well-known inhibition of auxin-stimulated growth, calcium salts would enhance gibberellin-stimulated growth and cytokinin-stimulated growth (Figure 1). The interactions were studied using oat coleoptile elongation, lettuce hypocotyl elongation, and *Xanthium* cotyledon enlargement, respectively, as assays. Leopold et al. also showed that calcium salts inhibited the actions of ethylene in the swelling of etiolated pea stems and enhanced the inhibitory effects of abscisic acid on the germination of lettuce seeds (Figure 2). They concluded that calcium salts were capable of altering the effectiveness of the plant hormones.

In none of the cases cited was there unequivocal evidence that the salt action was specific to the hormone action; in each case it was possible that the salt would have had effects independent of the hormonal effects. In order to distinguish between calcium salt effects independent of the hormone and effects directly involved with the hormone, tests were carried out using rice mesocotyls which are stimulated to elongate by either of two hormones; the calcium effects were compared in the presence and absence of each of these hormones. The results are illustrated in Figure 3; ethylene itself stimulates mesocotyl elongation, and $CaCl_2$ inhibits that ethylene stimulation (the $CaCl_2$ had no appreciable effect itself). In parallel experiments with the same type of tissue, gibberellin stimulated elongation, and $CaCl_2$ markedly enhanced the gibberellin stimulation. This implies that not only can calcium salts alter some regulatory effects of each of the plant hormones but that the calcium effects can be opposite in sign if growth elongation is being stimulated by different hormones.

If the salt effects were related to the destabilization effects of calcium, one would expect a range of effects with various salts of the Hofmeister

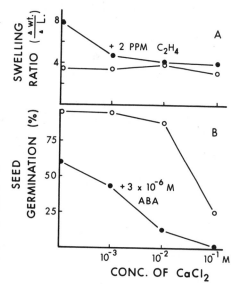

Figure 2.   Effects of CaCl₂ on bio-assays for ethylene and abscisic acid. A: effects on swelling of etiolated pea stems in the presence and absence of ethylene and B: effects on germination of lettuce seeds in the presence and absence of abscisic acid (13).

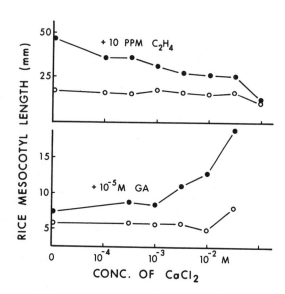

Figure 3.   Alterations of the growth of rice mesocotyls by CaCl₂ in the presence or absence of either ethylene or gibberellic acid (13)

series, with salts of the strongest stabilization effects having quite different effects from those of the destabilizer, calcium. The effectiveness of
various salts in altering auxin actions on growth has been compared
(*13, 14*). Ammonium salts are capable of enhancing the auxin effect, as
illustrated in Figure 4; salts which are intermediate in the Hofmeister
series are generally intermediate between the inhibitory effects of $CaCl_2$
and the promotive effects of $(NH_4)_2SO_4$. In addition to the salt effects
on auxin-stimulated growth, a contrast between calcium and ammonium
salts can be seen in the regulation of apical dominance in soybeans as

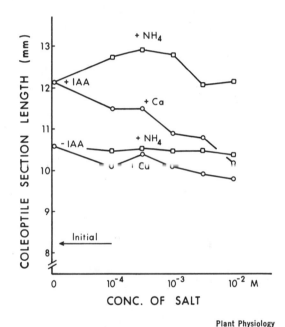

Plant Physiology

*Figure 4.   Effects of $(NH_4)_2SO_4$ and $CaCl_2$ on
the elongation of corn coleoptile sections in the
presence and absence of indoleactic acid (14)*

illustrated in Figure 5; calcium promotes the development of lateral
shoots in the presence of a cytokinin, and ammonium inhibits such
development. Again, in the case of abscission, calcium inhibits ethylene-
induced abscission (*8*), but ammonium sulfate promotes it (*14*). In the
regulation of leaf senescence, calcium salts can inhibit the development
of senescence in the presence of either a cytokinin or gibberellin (*7*). An
opposite effect is found for ammonium sulfate in this regulatory function,
and, as shown in Figure 6, the promotion of senescence by ammonium
sulfate can be reversed by $CaCl_2$. As indicated in the lower part of

Figure 6, the effects of calcium and ammonium salts on senescence have been linked with effects on the permeability of membranes (15).

An attractive possibility for explaining the means by which inorganic salts such as calcium and ammonium might alter hormonal regulation is as follows. The salts may alter the configurations of membranes or of macromolecules in such a way that the affinity of the hormone for its site of attachment is changed. Among the plant hormones, experimental techniques for measuring the attachment to a stereospecific binding site have been developed only for auxin (1). Experiments with pelleted membrane particles from corn coleoptiles have been done using naphthaleneacetic acid as the auxin and measuring the effects of $CaCl_2$ on the specific binding of the auxin. The specific binding is measured as the

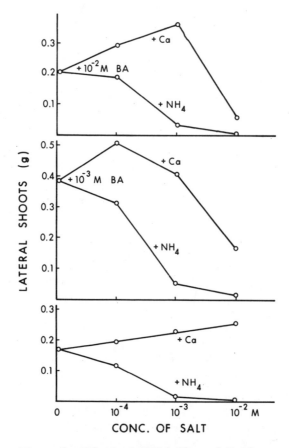

Figure 5.   Effects of $(NH_4)_2SO_4$ and $CaCl_2$ on the development of lateral shoots on soybean cuttings in the presence of zero, 10^{-3}, or $10^{-2}M$ benzyladenine

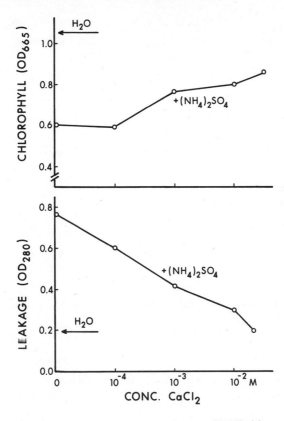

Figure 6.   Effects of $(NH_4)_2SO_4$ and $CaCl_2$ on the senescence of corn leaf discs, showing an acceleration of senescence (as a lowering of chlorophyll content and an increase in leakage of solutes out of the leaf sections) by $(NH_4)_2SO_4$ and a reversal of that effect by the addition of $CaCl_2$ (15)

difference between the amount of auxin radioactivity that spins down with the pellet in the presence of radioactive auxin alone and in the presence of radioactive auxin plus an excess of nonradioactive auxin, which could displace the radioactive molecules from specific binding sites. Figure 7 illustrates such experiments done in the presence of different concentrations of $CaCl_2$; it is evident that the $CaCl_2$ markedly increases the specific binding of the auxin to the materials in the pellet. The involvement of this binding site in the regulatory activities of auxin is not clear, and neither is it clear that the increase in auxin binding caused by the calcium is related to the inhibitory actions of calcium on auxin action. The deduction can be made, nevertheless, that $CaCl_2$ does

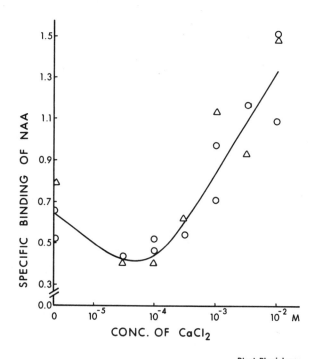

Figure 7.   *Effects of CaCl$_2$ on the binding of naphthaleneacetic acid to membrane pieces from corn coleoptiles* (15)

alter the specific binding of a plant hormone to a natural binding site in coleoptile cells.

These various data show that inorganic salts alter the actions of each of the known plant hormones, that instances exist in which salts from opposite ends of the Hofmeister series (or the distabilization/stabilization series) have opposite effects from one another, and that the characteristics of membranes are altered so that the specific binding of a hormone is changed by the presence of the salt.

The natural responsiveness of plant tissues to hormones may vary with changes in the configuration of action sites in the plant cells. Also, the art of regulating plant performance by exogenous substances may involve the solute microenvironment in the plant cells and their modification of presumed receptor sites for plant regulatory substances. It is reasonable to believe that at least some plant regulators operate by altering the specific binding of endogenous hormones (*16, 17*). Perhaps a further range of regulatory possibilities exists by exploitation of either inorganic or organic chemicals which can alter the destabilization/stabilization features of membranes and macromolecules in the plant and thus presumably alter the effectiveness of either endogenous or exgenous plant regulators.

Literature Cited

1. Hertel, R., Thomson, K. S., Russo, V. E. A., *Planta* (1972) **107**, 325.
2. Hofmeister, F., *Exp. Pathol. Pharmacol.* (1888) **24**, 247.
3. von Hippel, P. H., Schleich, T., "Structure and Stability of Biological Macromolecules," Timashev, S. N., Fasman, G. D., Eds., Marcel Dekker, Inc., New York, 1969, pp. 417–573.
4. Cooil, B. J., Bonner, J., *Planta* (1956) **48**, 696.
5. Thimann, K. V., Schneider, C. L., *Am. J. Bot.* (1938) **24**, 270.
6. Burstrom, H., *Plant Physiol.* (1954) **7**, 332.
7. Poovaiah, B. W., Leopold, A. C., *Plant Physiol.* (1973) **52**, 235.
8. Ibid., **51**, 848.
9. Bangerth, F., *Planta* (1974) **117**, 329.
10. Lau, O. L., Yang, S. F., *Planta* (1974) **118**, 1.
11. Posner, H. B., *Plant Physiol.* (1969) **44**, 562.
12. Hillman, W. S., Posner, H. B., *Plant Physiol.* (1971) **47**, 586.
13. Leopold, A. C., Poovaiah, B. W., dela Fuente, R. K., Williams, R. J., "Plant Growth Substances, 1973," Hirokawa Publishing Co., Tokyo, 1974, pp. 780–788.
14. Poovaiah, B. W., Leopold, A. C., *Plant Physiol.* (1976) **58**, 783–785.
15. Ibid., 182–185.
16. Thomson, K. S., "Hormonal Regulation of Plant Growth and Development," Kaldewey, H., Vardar, Y., Eds., Verlag Chemie, Weinheim, 1972, pp. 83–88.
17. Thomson, K. S., Leopold, A. C., *Planta* (1974) **115**, 259.

RECEIVED September 22, 1976.

5

Management of the Cotton Plant with Ethylene and Other Growth Regulators

PACE W. MORGAN

Department of Plant Sciences, Texas Agricultural Experiment Station, Texas A&M University, College Station, Tex. 77843

The complexity and cost of production methods in cotton culture are increasing the need for and the feasibility of effective plant growth regulators to improve managerial control of the crop. Response goals include: (1) improved seedling vigor, (2) early flowering, (3) promotion of fruiting under conditions that favor excessive vegetative growth, (4) improved fruit set and translocation to fruit, (5) improved fiber and seed quality and yield, (6) early termination of flowering, and (7) improved harvest-aid systems. Current research centers on seedling vigor, plant height, flowering, fruit set, and early termination of flowering and growth. Manipulation of the ethylene physiology of plants can influence many responses, some of which are relevant to cotton production. Most other growth regulators and some plant hormones modify ethylene physiology, resulting both in problems and in opportunities to discover unique responses.

Cotton as a profitable crop is uniquely sensitive to competition from synthetic substitutes—the man-made fibers. For this and other reasons there is a particularly pressing need in the cotton production business for new ways to cut costs or to increase efficiency. Plant growth regulators may be a major avenue by which unit production costs can be reduced. It seems likely that systems for chemical manipulation of crop performance would be readily accepted by cotton producers.

Recent changes in crop production systems are increasing the feasibility of using chemical growth regulants. Cotton is experiencing many of these important changes. For example, the use of high-density populations of dwarf-type compact plants and the more conservative uses of

42

irrigation water and fertilizer will produce uniform plants in similar growth stages with a compressed fruiting period. This condition is more compatible with chemical managment than an extended, indeterminate fruiting mode on large plants which results in a fruit load of variable age and location. A shorter time interval from planting to harvest will increase the need for uniform fruit maturity, efficient removal of leaves, and effective regrowth control to facilitate mechanical harvesting.

This review is written from the position that both the potential for successful development of growth regulator applications in cotton production and the need for these applications have increased significantly in recent years. The purpose is: (a) to analyze the types of responses which would be beneficial, (b) to review recent studies, (c) to point out the underlying importance of ethylene physiology to growth regulator investigations, and (d) to present some potentials of ethylene as an agricultural chemical. Ethylene physiology (*1, 2, 3*) and the manipulation of ethylene as a strategy in agricultural production (*4, 5*) have been reviewed recently in detail.

Desirable Plant Responses

How would the commercial producer manipulate the cotton plant if he had his choice? An analysis of the possible responses and further consideration of their potential value should aid in the development of research objectives. That approach is used here.

Cotton is a tropical deciduous perennial from a Mediterranean climate which is cultured as an annual. It is produced under divergent environmental conditions which result in great differences in cultural systems and management. For example, it is produced on dry land under marginal levels of rainfall, in non-irrigated areas with usually abundant rainfall (rainbelt), in areas with supplemental irrigation, and in arid regions with complete irrigation. In addition, both conventional (40-in.) and narrow (10-inch) row spacing and high and low plant densities are now being used. The plant responses one would desire to control in a certain culture system might be of little value in another. The following analysis of the crop considers the major differences in how the crop develops and how it is managed.

Seed Germination. Across the American cotton belt, cotton germinates under environmental stresses such as low soil temperature, moisture deficit or excess, oxygen deficits, and mineral extremes. Seed treatments to increase tolerance to early season stresses would be very beneficial. When large scale replanting depletes the best seed, lower quality seed must be used. So any treatment which would improve the performance of such seed, even under favorable environmental conditions, would be valuable. In the future, plant breeders may select breeding lines for those seed quality characteristics exclusively related to food value (resistance to weathering, etc.). If hard seed coat, physiological dormancy, or some other agronomically undesirable characteristic accompanied such genetic modification of the plant, it might be feasible to circumvent those characteristics chemically in planting seed. In doing so, one would be able to realize the benefits of increased food quality from that portion of the cottonseed which is processed.

During the seedling and early vegetative stage, cotton often is subjected to cold weather; being of tropical origin, the plant suffers chilling injury which has long term growth and flowering effects (6, 7). Presently there is interest in substances to prevent or circumvent chilling injury, and such could be useful to cotton producers.

Vegetative Development–Flower Initiation. In the past, an accepted strategy for improving cotton yields was to increase early fruit set. That is no longer a complete or adequately specific solution. This goal is being reached, in a sense, in the narrow row or high density planting systems (8, 9). Plant competition forces earlier flowering and although individual plants have fewer bolls, the total acre yield can be more. Further, there are some records of increased fruit set without increased yield (10, 11); thus, in some circumstances fruit load may not limit yield. There are, however, several processes of the cotton plant during the pre-fruiting period which might be manipulated in a beneficial way, depending upon the nature of the cultural system involved. A major need is to prevent excessive vegetative growth in situations of luxuriant moisture and fertility. Managers have learned to moisture stress cotton marginally in the desert-irrigated regions to shift from vegetative to reproductive growth, but the problem is not limited to those areas. A mild chemical stress agent would be very useful; it should favor reproductive over vegetative growth while possibly shortening and strengthening the main stem. Vigorous early growth and fruit set are drought- and disease-escape mechanisms in drier, non-irrigated areas; however, they are unachieved because high density planting is not feasible under strong moisture limitations. In such areas a chemical which would set more fruit earlier could reduce the frequency of crop failures.

Fruiting. Once flowering starts, it is generally recognized that the crop should be set as quickly as possible. A regulator to improve fruit set could be useful, for example, to offset the effect of stresses such as heat, moderate moisture lack, minor insect damage, and cloudy weather. On the other hand, a better goal might be for a regulator to hasten resumption of flowering and fruit setting after the relief of stress.

Fruit Growth. Perhaps the most obvious growth regulator goal in cotton culture is to discover a way to promote fiber production chemically. This might be accomplished by: (a) improving mobilization of photosynthate in order to promote the rate of fiber elongation and secondary cell wall deposition, (b) prolonging the duration of fiber development, or (c) promoting initiation of a greater number of fibers per seed. New information on the interplay between elongation and secondary cell wall growth (*12, 13*) and the hormonal regulation of fiber development in tissue culture (*14, 15*) may aid efforts to develop longer, stronger fibers in the boll. There is evidence suggesting that photosynthesis alone does not limit yield. Perhaps mobilization of photosynthate to the boll is the key; if so, inhibition of translocation to and storage of starch in the stem and roots might improve translocation as effectively as a direct promotion of the mobilizing strength of the fruit.

Possibly one of the most interesting and potentially beneficial concepts now being explored is early termination of flowering—once achieved genetically in the strongly determinate varieties. The practice comes to our attention at this time primarily because of insect problems; once the main crop is set, if further flowering could be terminated, substantial savings in insecticides and other costs could be realized. This system has been demonstrated to be economically sound and feasible in Arizona where termination can be initiated by withholding water (*16, 17*). It seems that neither the use of highly determinate varieties nor the withholding of irrigation water is a generally satisfactory early termination system. Basically, the objective is a chemical regulator which will quickly bring fruiting meristems into dormancy or inactivity while not impairing the plant's ability to channel most of the available energy into the fruit already present.

Harvest. With defoliation or chemical desiccation almost a universal practice and with economical, effective materials in established markets, this would appear to be one phase of the crop which can be ignored. Such is not the case. One persistent problem is regrowth after defoliation when rain or other weather delays harvesting. A good regrowth inhibitor to go along with the basic harvest-aid chemical would be useful on large acreages. Use of arsenicals may not always be permitted on stripper-harvested cotton, and an acceptable substitute is not apparent. Additional acres are harvested by stripper harvesters each year, and high-density, narrow-row culture systems may accelerate this trend. The harvest phase of cotton production could benefit from some new products.

Research Areas

Seed Treatments. Several efforts to improve seed germination chemically have been made in recent years. The inhibition of lateral root development in cotton seedlings which can result from exposure to the herbicide trifluralin have been relieved with IAA or kinetin (*18*). Soil-incorporated D-α-tocopherol or oleic acid and cottonseed oil also reduced trifuralin damage (*19*). Many workers have reported that multiple treatments of cotton seeds with insecticides and fungicides can produce detrimental interactions (*20, 21, 22*), and any substance which would prevent this type of pesticide damage could be a valuable component of a seed treatment package.

Chilling injury to Pima S-4 cotton seedlings during germination was reduced by preconditioning seed before planting with either water alone, aqueous GA, or cyclic AMP (*23*). Germination of cottonseed was inhibited by ABA; furthermore, ethephon, GA_3, and kinetin all partially overcame the inhibition by ABA (*24*). No desirable result was recognized from soaking seeds in NAA prior to planting, and the highest concentrations tested reduced stands and delayed maturity (*25*). Gibberellic acid applied to cottonseeds promoted lipase activity. Presumably the endogenous gibberellins do the same (*26*). Aflatoxins inhibited germination and lipase activity (*26*).

Oil seeds, cotton, and peanuts, in particular, produce ethylene at rather high rates during and immediately following germination (*27*). High germinability and seedling vigor are correlated with high ethylene production during germination (*27*); however, there is as yet no published evidence that ethylene or ethephon will improve germination of cottonseeds.

Flowering, Fruit Set, and General Plant Development. Much of the current research on cotton involves applications of materials before initiation or during flowering to modify fruiting. One recognized problem is that the insecticide, methyl parathion, delays the maturity of cotton and reduces yields, at least under some environmental conditions, by raising the nodal position of the first fruiting branch (*28, 29, 30*). The delay can further complicate late season insect control problems and illustrates that there may be a potential for improvement in cotton by making the crop mature earlier. During extended, very faborable growing seasons, however, methyl parathion does not delay maturity and actually increases yields (*31, 32, 33*). Similar inconsistency can be expected in responses sought for beneficial results. Practices may be possible for specific areas and culture systems which will not be effective worldwide. On some sites, Pima S-4 cotton sets its first fruit too low for efficient mechanical picking, and lower flower buds could be removed with ethephon. The node level of the first fruit was raised about four nodes (*34*).

The problem of cotton "going vegetative" can result from high soil fertility, abundant moisture, loss of early flowers to insects, or combinations of these factors. The result is usually rank vegetative growth and low yields. Even if fruit set is acceptable, excessively tall cotton can lodge, be subject to excess boll rot, and be difficult to harvest. Currently there is interest in controlling height and vegetative growth. In cases where little fruit set would occur without this practice, yield would hopefully be increased. However, in other cases where average yields would occur, it might be acceptable if the regulator simply reduces height and does not lower yields. The growth retardants, CCC (2-chloroethyltrimethylammonium chloride) and CMH (N-dimethyl-N-β-chloroethylhydrazonium chloride), were used on Acala cultivars in Israel and reduced that height growth which occurred after application as much as 50% without reducing yield (35). Applications were made during the first week of flowering, and the results were improvements over earlier efforts cited by the authors. African Upland variety SATU 65 in Uganda responded to CCC with a reduction in both yield and height (36). Growth inhibitors BAS 0660W (dimethylmorpholinum chloride) and BAS 0640W [dimethyl-N-(β-chloroethyl)hydronium chloride] effectively reduced plant height and increased earliness and ginning percentage without damaging fiber quality, but the effect was less on irrigated than on rain-grown cotton (37). The most serious lodging problem occurs with the irrigated cotton (37).

One of the early efforts to limit height of the cotton plant with CCC was in Mississippi where undesirable yield reductions occurred (38); however, in recent experiments, Thomas was able to reduce height 16 in. with CCC without a significant reduction in yield (39). Unsuccessful efforts have been made to modify cotton yield with sub-lethal applications of simazine (S-triazine) and terbacil (3-tert-butyl-5-chloro-6-methyluracil) (40). Freytag and Wendt (41) found that soil-injected ethylene increased yields of cotton in some but not all moisture-tension regimes. Reduced height was obtained with TIBA (2,3,5-triiodobenzoic acid), but it lowered the yield (42).

Some effects of growth regulators on fiber properties have been studied. An increased yield of one Indian cotton, MCU-1, was achieved with CCC, but it had no effect or reduced yields of three other varieties (43). Some abnormal bolls were produced, and the lint from these bolls was coarser and stronger. Bhatt et al. (44) studied the effects of several growth regulators on fiber quality in India. NAA increased fiber fineness at a low concentration and had the reverse effect at higher ones. IAA improved length slightly and improved fineness. SADH (1,1-dimethylaminosuccinic acid) increased length and Phosfon (2,4-dichlorobenzyltributylphosphonium chloride) increased fineness. At a lower concentration, tested CCC increased fiber coarseness, but higher concentrations increased length and fineness while decreasing strength, maturity, and yield. Gibberellic acid applied to the Indian variety PRS-72 did not reduce seed cotton yield but increased fiber length significantly (45).

On Laxmi cotton in India, GA_3 did not increase fiber length or change other fiber characteristics (46). Studies on natural growth substances on cotton fibers suggest that there may be specific fiber elongation hormones (47), and thus, failure to achieve a major, consistent improvement in fiber properties to date may simply be the result of the failure to test the right compound. For that reason, the potential for chemically improving fiber length should not be ignored.

Early Termination. Work is underway to develop a system for chemical termination of flowering. Thomas (39) applied CCC and DPX-1840 [3,3a-dihydro-2-(p-methoxpyhenyl)-8H-pyrazolo[5,1-a]isoindol-8-one] in order to retard late season flowering and found the latter material more effective. Additional studies revealed that DPX-1840 was readily absorbed and translocated to stem tips where it eventually retarded growth without serious effects on flower or boll development (48). In Arizona, studies were undertaken to reduce food (cotton flower buds) for the pink bollworm larvae and thus reduce the population going into diapause (49). Results indicated that 2,4-D (2,4-dichlorophenoxyacetic acid), CCC, chlorflurenol (methyl-2-chloro-9-hydroxyfluorene-9-carboxylate), Pennwalt TD 1123 (3,4-dichloroisothiazole), and BasF 0660 (N-dimethylmorpholine chloride) showed promise. Ethephon, DPX-1840, cacodylic acid (hydroxydimethylarsine oxide), Sustar 2-S MM [3'-(trifluoromomethylsulfonamido)-p-acetotoluidide], and Dalapon were not promising. Treatments which reduced green bolls remaining at harvest also reduced the number of pink bollworm larvae. This work was extended in 1974 with several mixtures of the above compounds acting very effectively; the most effective termination treatments reduced larvae 93–97% with less than 5% yield reduction (50). Although 2,4-D is apparently the most effective compound tested, its translocation into seed and very high damage potential on seedlings makes its use questionable. Thus, additional work is needed to perfect an acceptable early termination system.

Defoliation. Interest in defoliation has been low in recent years. One relatively new development is the "wiltant" which is applied only shortly before harvest (51). As an outgrowth of some basic studies, several auxin transport inhibitors, TIBA, DPX-1840, Alanap (N-1-naphthylphthalamate), and morphactins (2-chloro-9-hydroxyfluorene-9-carboxylic acid), were shown to promote ethylene- and ethephon-mediated leaf abscission (52, 53). Subsequently, GA_3 was found to be even more active in promotion of ethylene-induced abscission (54). It now appears that the GA_3 counteracts the inhibitory effect of auxin on ethylene-induced leaf abscission (55); thus, GA_3 might improve the performance of any defoliant that achieves part of its action by stimulating stress-induced ethylene production and lowering the natural auxin content of the damaged leaves. From a slightly different approach, Sterrett et al. (*56*) have shown a synergistic effect of ethephon and the defoliant endothall (7-oxabicyclo[2,2,1]heptane-2,3-dicarboxylic acid) at levels at which neither is effective alone. Field work is currently underway to condition cotton for defoliation (*57*). Miller (*58*) has devised a new technique for improving the efficiency of desiccants which involves application to the stem so that the chemical enters the transpiration stream and moves internally to the leaf.

Ethylene–Ethephon: Potential in Growth Regulation

In recent years ethylene has enjoyed a rather glamorous state among the plant hormones. It has been considered in a relative flood of papers and has been implicated in a wide array of plant processes. Ethylene is also enjoying a growing status as a practical plant growth regulator. This probably stems from three major reasons: (a) ethylene has relatively minor residue problems, (b) ethylene is involved in many naturally regulated plant processes, and (c) the number of options for regulating ethylene physiology is large and increasing (*4, 5*).

Ethylene promotes a large array of responses in seeds, plants, and fruits (Table I), some similar to and some dissimilar to effects of auxins. Those with actual or potential commercial application include: (a) growth promotion of seedlings (rice), (b) inhibition of height growth, (c) root initiation, (d) chlorophyll destruction (citrus), (e) flower initiation (pineapple, bromeliads), (f) stimulation of fruit growth (figs),

Table I.   Effects of Applied Ethylene on Plants and Plant Parts Compared with Similar and Dissimilar Effects of Auxins (*5, 67*)

Similar Plant Responses to Ethylene and Auxins

Growth inhibition	Pigment synthesis inhibition
Growth promotion	Flower initiation
Geotropism modification	Flower inhibition
Tissue proliferation	Flower sex shifts
Root and root hair initiation	Fruit growth stimulation
Leaf epinasty	Fruit degreening
Leaf movement inhibition	Fruit ripening
Formative growth and hook formation	Respiratory changes
Chlorophyll destruction	Storage product hydrolysis
Pigment synthesis promotion	Latex secretion promotion

Dissimilar Plant Responses to Ethylene and Auxins

Abscission and dehiscence
Seed and bud dormancy release
Apical dominance release

(g) fruit ripening and loosening, (h) promotion of storage material hydrolysis, (i) promotion of secretion (rubber), (j) leaf, fruit, and flower abscission, (k) release of lateral buds from apical dominance, and (l) release of seed and bud dormancy. The potential is limited because some effects are known in only isolated species noted in parentheses above. Also, desirable and undesirable responses may occur simultaneously. On the other hand, much work with ethylene has been done with high concentrations or long durations which may have obscured desirable responses. A range of ideas has been considered relative to the use of ethylene in cotton, including: (a) improving seedling vigor, (b) enhancing early-season branching and flowering, (c) termination of flowering, (d) defoliation and (e) boll opening (4). None of these uses has been achieved on a commercial scale.

Ethylene physiology of the plant can be manipulated in a variety of ways. In the past, the use of ethylene was limited to exposure of plants to the gas in containers; thus, field applications were impractical. This limitation was removed by the discovery and commercial development of ethephon in which the liquid active ingredient, 2-chloroethyl phosphonic acid, is converted to ethylene by the plant (59). Other means of modifying ethylene physiology have been recognized and discussed (4, 5). It is possible to stimulate ethylene synthesis with auxins (60, 61, 62, 63), abscisic acid (64), defoliants (65), ascorbic acid (66), cycloheximide (66), and iron salts (66), among other compounds. A number of physical, environmental, microbial, and insect stresses increase ethylene synthesis, including moisture stress (67) and air pollutants (68).

In some cases it might be desirable to inhibit ethylene synthesis chemically to prevent responses mediated by naturally produced ethylene or stress-produced ethylene. Although some substances do inhibit ethylene production modestly—e.g. TIBA (69)—no outstanding regulator of this nature has been discovered. Another possibility is to promote or inhibit ethylene action. Promotion can be accomplished by auxin transport inhibitors and GA in cases where auxins and ethylene have opposite effects (52, 53, 54, 55). Recently, silver ion was found to be a potent inhibitor of ethylene action (70). Ethylene action also can be inhibited by lowering the temperature and O_2 level or increasing the CO_2 level (2). These manipulations are not usually practical in the field. An effective inhibitor of ethylene action might be a useful growth regulator. Removal of ethylene significantly delays natural ethylene responses such as fruit ripening (71), but the procedure requires putting the plant or fruit in a container in which a partial vacuum can be sustained. It is thus only applicable to harvested fruit and vegetables, potted plants, and similar items.

Ethylene Physiology in All Growth Regulator Applications

Since many growth regulators modify ethylene synthesis and responses to ethylene (2, 72), including many of the substances listed in the previous section, it is possible that some failures to achieve desired responses in past studies could have been the result of secondary effects on the ethylene physiology of the plant. Some of these problems might be alleviated when their causes are recognized. It seems desirable, then, to know how new growth regulators modify ethylene physiology in cotton prior to field testing. Stress effects on hormone physiology also need more attention in the future. Some of the inconsistency of the effects of several growth regulators on fruit set, for example, may result from differences in environmental stresses immediately before or at the time of the treatment.

The attention focused on ethylene physiology in this section has another message. Perhaps this is a case where basic understanding of hormonal–environmental interactions can contribute, in the long run, to improved agricultural practices. While there is still no way to custom design a regulator to do a certain job, we are learning more about the secondary effects and interactions which must be considered. Ethylene, both ubiquitously present and uniquely phytoactive, seems to be critical to many goals of growth regulator technology.

Literature Cited

1. Abeles, F. B., *Plant Physiol.* (1972) **23**, 259–92.
2. Abeles, F. B., "Ethylene in Plant Biology," Academic, New York, 1973.
3. Yang, S. F., "Recent Advances in Phytochemistry," B. C. Runeckles et al., Eds., **7**, 131–164, Academic, New York, 1974.
4. Morgan, P. W., *Proc. 1973 Beltwide Cotton Prod. Res. Conf.* (1973), 95–100 (National Cotton Council, Memphis, Tenn.).
5. Morgan, P. W., *Misc. Publs. Tex. Agri. Exp. Stn. 1018* (1972), 12.
6. Christiansen, M. N., *Plant Physiol.* (1967) **42**, 431–433.
7. Powell, R. D., Amin, J. V., *Cotton Grow. Rev.* (1969) **46**, 21–28.
8. Ray, L. L., *Proc. 1970 Beltwide Cotton Prod. Res. Conf.* (1970) 57.
9. Niles, G. A., *Proc. 1970 Beltwide Cotton Prod. Res. Conf.* (1970), 63–64.
10. Ergle, D. R., McIlrath, W. J., *Bot. Gaz.* (1952) **114**, 114–122.
11. Walhood, V. T., *Proc. Cotton Defoliation and Physiol. Conf., 12th* (1957) 24–30. (National Cotton Council, Memphis, Tenn.).
12. Schubert, A. M., Benedict, C. R., Berlin, J. D., Kohel, R. J., *Crop Sci.* (1973) **13**, 704–709.
13. Schubert, A. M., Benedict, C. R., Gates, C. E., Kohel, R. J., *Crop Sci.* (1976) **16**, 539–543.
14. Beasley, C. A., *Science* (1973) **179**, 1003–1005.
15. Beasley, C. A., Egli, E., *Proc. 1976 Beltwide Cotton Prod. Res. Conf.* (1976) 39–44.
16. Willett, G. S., Taylor, B. B., Buxton, D. R., *Proc. 1973 Beltwide Cotton Prod. Res. Conf.* (1973) 37–39.
17. Payne, H. L., Buxton, D. R. Hathorn, S., Jr., Briggs, R. E., Patterson, L. L., *Proc. 1976 Beltwide Cotton Prod. Res. Conf.* (1976) 71–74.

18. Hassawy, G. S., Hamilton, K. C., *Weed Sci.* (1971) **19**, 256–268.
19. Christiansen, M. N., Hilton, J. L., *Crop Sci.* (1974) **14**, 489–490.
20. Ranney, C. D., *Proc. Meet. Cotton Disease Council, 24th* (1964) 59–63 (National Cotton Council, Memphis, Tenn.).
21. Ranney, C. D., *Crop Sci.* (1972) **12**, 346–350.
22. Milton, E. B., *Crop Sci.* (1972) **12**, 93–94.
23. Cole, D. F., Wheeler, J. E., *Crop Sci.* (1974) **14**, 451–454.
24. Halloin, J. M., *Proc. 1974 Beltwide Cotton Prod. Res. Conf.* (1974) 36.
25. Coats, G. E., *Proc. 1967 Beltwide Cotton Prod. Res. Conf.* (1967) 134–137.
26. Jones, H. C., *Nature* (1967) **214**, 171–172.
27. Ketring, D. L., Morgan, P. W., Powell, R. D., "Plant Growth Substances 1973," Y. Sumiki, Ed., 891–899, Hirokawa Publ. Co., Inc., Tokyo, 1974.
28. Brown, L. C., Cathey, G. W., Lincoln, C., *J. Econ. Entomol.* (1962) **55**, 298–301.
29. Roark, B., Pfrimmer, T. R., Merkl, M. E., *Crop Sci.* (1963) **3**, 338–341.
30. Beasley, C. A., *Proc. 1969 Beltwide Cotton Prod. Res. Conf.* (1969) 99–102.
31. Weaver, J. B., Jr., Harvey, L., *Proc. Cotton Improvement Conf., 15th,* (1963) 91–104 (National Cotton Council, Memphis, Tenn.).
32. Weaver, J. B., Jr., *Proc. 1973 Beltwide Cotton Prod. Res. Conf.* (1973) 73.
33. Horst, B. J., *Proc. 1973 Beltwide Cotton Prod. Res. Conf.* (1973) 74–75.
34. Pinkas, L. L. H., *Crop Sci.* (1972) **12**, 465–466.
35. Marani, A., Zur, M., Eshel, A., Zimmerman, H., Cameli, R., Karadavid, B., *Crop Sci.* (1973) **13**, 429–432.
36. De Silva, W. H., *Cotton Grow. Rev.* (1971) **48**, 131–135.
37. Follin, J. C., *Coton Fibres Trop. Engl. Ed.* (1973) **28**, 449–451.
38. Thomas, R. O., *Crop Sci.* (1964) **4**, 403–406.
39. Thomas, R. O., *Proc. 1972 Beltwide Cotton Prod. Res. Conf.* (1972) 49.
40. Smith, D. T., *Agron. J.* (1972) **63**, 945–947.
41. Freytag, A. H., Wendt, C. W., *Agron. J.* (1972) **64**, 524–526.
42. Thomas, R. O., *Proc. 1967 Beltwide Cotton Prod. Res. Conf.* (1967) 222–226.
43. Bhatt, J. G., Ramanujam, T., *Cotton Grow. Rev.* (1972) **49**, 354–358.
44. Bhatt, J. G., Raman, C. V., Sankaranarayanan, T. G., Iyer, S. K., *Cotton Grow. Rev.* (1972) **49**, 160–165.
45. Bhatt, J. G., Ramanujam, T., *Cotton Grow. Rev.* (1971) **48**, 136–139.
46. Sitaram, M. S., Abraham, E. S., *Cotton Grow. Rev.* (1973) **50**, 150–151.
47. Mitchell, J. W., York, G. D., Worley, J. F., *Plant Physiol.* (1966) **41** (suppl.), 1x.
48. Thomas, R. O., Hacskaylo, J., *Proc. 1973 Beltwide Cotton Prod. Res. Conf.* (1973) 36.
49. Kittock, D. L., Arle, H. F., Bariola, L. A., *Proc. 1974 Beltwide Cotton Prod. Res. Conf.* (1974) **71**, 55–56.
50. Kittock, D. L., Arle, H. F., Bariola, L. A., *Proc. 1974 Beltwide Cotton Prod. Res. Conf.* (1974) **71**.
51. Edwards, G. C., Rutkowski, A. J., Schwendinger, R. B., Canzano, P. S., *Proc. 1968 Beltwide Cotton Prod. Res. Conf.* (1968) 62–68.
52. Morgan, P. W., Durham, J. I., *Plant Physiol.* (1972) **50**, 313–318.
53. Morgan, P. W., Durham, J. I., *Planta* (1973) **110**, 91–93.
54. Morgan, P. W., Durham, J. I., *Plant Physiol.* (1975) **55**, 308–311.
55. Morgan, P. W., *Planta* (1976) **129**, 275–276.
56. Sterrett, J. P., Leather, G. R., Tozer, W. E., *Plant Physiol.* (1973) **42**, 1021–1022.
57. Arle, H. F., *Proc. 1976 Beltwide Cotton Prod. Res. Conf.* (1976) 49.
58. Miller, C. S., *Proc. 1975 Beltwide Cotton Prod. Res. Conf.* (1975) 63.
59. de Wilde, R. C., *Hort. Sci.* (1971) **6**, 364–370.
60. Zimmerman, P. W., Wilcoxon, F., *Contrib. Boyce Thompson Inst.* (1935) **7**, 209–229.

61. Morgan, P. W., Hall, W. C., *Physiol. Plant.* (1962) **15**, 420–427.
62. Morgan, P. W., Hall, W. C., *Nature* (1964) **201**, 91.
63. Burg, S. P., Burg, E. A., *Proc. Nat. Acad. Sci. USA* (1966) **55**, 262–269.
64. Cracker, L. E., Abeles, F. B., *Plant Physiol.* (1969) **44**, 1144–1149.
65. Hall, W. C., *Bot. Gaz.* (1952) **113**, 310–322.
66. Cooper, W. C., Rasmussen, G. K., Rodgers, B. J., Reece, P. C., Henry, W. H., *Plant Physiol.* (1968) **43**, 1560–1576.
67. McMichael, B. L., Jordan, W. R., Powell, R. D., *Plant Physiol.* (1972) **49**, 658–660.
68. Abeles, A. L., Abeles, F. B., *Plant Physiol.* (1972) **50**, 496–498.
69. Abeles, F. B., Rubinstein, B., *Plant Physiol.* (1964) **39**, 963–969.
70. Beyer, E. M., Jr., *Plant Physiol.* (1976) **58**, 268–271.
71. Burg, S. P., Burg, E. A., *Science* (1966) **153**, 314–315.
72. Morgan, P. W., *Acta Hortic.* (1973) **34**, 41–54.

RECEIVED October 29, 1976.

Economic Potential

6

Economic Potential of Plant Growth Regulators

DAVID T. MANNING

Research and Development Department, Union Carbide Corp.,
Agricultural Products Division, S. Charleston, W. Va. 25303

*While plant growth regulants (PGRs) have emerged as a
new, rapidly growing segment of agricultural chemicals,
the novelty, risks, and technological difficulties of this area
are challenges to be considered. Thus, many scientists seeking new experimental regulants for study in their specialty
areas raise questions as to the practical results forthcoming.
To this end better understanding is needed of how economic
values for various regulant uses can be estimated and then
translated into manufacturing incentives. Besides the discovery of new molecules, advances in PGRs will involve
novel uses of old compounds and will depend upon developments in plant breeding. The optimum development of
science in PGRs is seen as integrated with research in these
other areas of crop production.*

In considering the economic potential of plant growth regulants (PGRs)
some definitions of terminology are required. In one sense the title
topic refers to the future impact of growth regulant products on agriculture in terms of increased values of grower products. These values
will be reflected by a dollar volume of PGR sales as a segment of the
total future agricultural chemicals market. In another sense, PGR economic potential relates to the judgment of the economic values of new
PGR product candidates in individual crop use categories. Many persons
in industry, including those involved in marketing and in research and
product development, must make, and are guided by, these judgments
which determine the spectrum of uses for which a new compound is to
be developed and, ultimately, whether the new compound is carried
further or dropped.

In a 1971 review, Wittwer was optimistic about PGRs but noted that to that date their growth had been limited by a limited number of high volume demands (1). Today his optimism seems justified. We have seen strong growth in horticultural crop regulants and exciting developments in the areas of ethylene-releasing compounds and a variety of agents for increasing the sucrose content of sugarcane. In 1975 *Farm Chemicals* featured an article proclaiming "the threshold of the Age of PGRs" (2). The number of papers and patent issuings on the application of regulants to crop plants has increased rapidly each year since 1970. Still there are many questions and uncertainties for those joining this "gold rush." Screening for these agents is highly difficult, and field evaluation and proof of candidate efficacy are exquisitely sensitive to compound dose, timing, and interactions with the environment. Finally, these agricultural chemicals which function not by killing but by enhancing plant performance are spared none of the regulatory and economic hurdles faced in introducing new pesticides.

A successful PGR must be profitable both to the grower and to the manufacturer, or else it will remain a scientific curiosity. There have been disappointments and disillusionments to the grower in this relatively new area. University workers express eagerness to work with experimental compounds of promise in their specialty areas yet also are concerned with the practical results of their work. The high costs and the time required for introducing a new product are becoming widely appreciated. It would be of great benefit to these workers, and indeed to all participants in growth regulant technology, to have a better understanding of how economic values for various regulant uses can be estimated and translated into manufacturing incentives.

With sufficient assumptions made in relation to manufacturing cost, compound potency, possible market scope and share, and percent yield increase it is possible to project, at least within limits, the economic possibilities of a new plant growth regulant operating in areas having some precedent, such as for yield increases with soybeans, rice, or corn. Where PGR values are related directly to labor cost replacement or advantages in harvest timing, other estimation models will have to be used. Present topics of interest include economic aspects of PGRs in horticultural and agronomic crops and in floriculture and beet sugar production. PGRs are also effective in optimizing quality of fruit. With the quality aspect in mind we may wonder about the future likelihood of economic incentives for increasing the protein content of cereals, with growth-active chemicals supplementing plant breeding programs. This is a potential worldwide concern whose scope fits the multinational structure of the agricultural chemical industry. Perhaps some guidance can be obtained from recent plant breeding parallels.

Amid the flush of discovery of new chemical structures with growth activity, two other developments can be expected to influence this new technology strongly. First, much of value remains to be learned in the new uses of old agricultural chemicals. There is no more dramatic example of this than the success of certain herbicides and their derivatives in cane sugar enhancement. New discoveries can be expected from experimentation with growth-active substances in non-standard applications with other crops and with unconventional doses and timing. Modest improvements in crop performance, if consistent, can be quite valuable with compounds whose development costs have, in essence, already been paid.

A second point is that PGR workers must be aware of developments in genetics. There can be no reasonable future for chemists who are developing new regulant compounds and plant breeders who are developing improved cultivars in separate worlds. New techniques of plant modification by cell culture and hybridization portend a future of much more rapid plant improvements than have been possible with conventional breeding techniques. With both plant growth regulants and modern molecular genetics capable of removing the same biochemical constraints upon crop performance, the PGR researcher may imagine crops becoming so productive that PGR benefits become increasingly marginal. Basic research in PGRs may have to be coordinated with modern plant genetics. This is not to speak negatively of the economic potential of PGRs but to say that along with improved plant genetics and crop protection chemicals, PGRs will play an integrated, interdependent role in the future of plant production.

Literature Cited

1. Wittwer, S. H., *Outlook Agric.* (1971) **6** (5) 205-217.
2. "Entering the Age of Plant Growth Regulators," *Farm Chem.* (1975) **138** (3) 15-26.

RECEIVED September 30, 1976.

7

Economic Value of Growth Regulants in Horticulture

A. W. MITLEHNER

Uniroyal Chemical, Division of Uniroyal, Inc., Bethany, Conn. 06525

Plant growth regulators can improve the yield of some horticultural crops by increasing the number of flowers and fruits. Their real success, however, is correlated with their ability to modify quality and timing more effectively than other cultural alternatives. Another important reason for the successful development of plant growth regulants in horticulture is their ability to reduce labor costs associated with the production of high-value crops. Growth regulators which hasten or delay maturity allow greater labor flexibility at harvest. Concentrated maturity increases the efficiency of mechanical harvesters in once-over harvest procedures. Expensive hand-labor operations, which are often used to maintain the form and shape of horticultural plants, can also be reduced with plant growth regulators.

The development of plant growth regulators has been given a significant amount of attention during the past few years. Unlike the relative maturity of conventional pesticide products, however, growth regulants are just starting to attain the sophistication needed to make the contribution of which they are truly capable.

Plant growth regulators have been known for a long time. A long list of endogenous growth regulants, which predate the modern pesticide industry, could be easily prepared. Despite the effort put forth by the public research community, nothing of great commercial significance has happened in the marketplace with these compounds.

During the late 40s and 50s, commercial research was centered upon the identification of "cidal-type" products. At the same time, a few companies started to anticipate the day when there would be a substantial number of chemical tools available to control weeds, insects, and diseases.

This led to a broadening of research efforts, and programs were implemented to identify chemicals capable of manipulating the vegetative and reproductive growth of plants.

Early growth regulant screening was not a very sophisticated effort. Procedures used to evaluate this activity needed development. More often than not, the screen consisted of an observation made in conjunction with applications made for other purposes. The significance of this cannot be appreciated unless one realizes that a "very interesting" response is valueless in a commercial research program unless there is a commitment to doing something with it. A short plant in a herbicide screen means that the compound is not a herbicide. Blossoms falling off a tomato plant in an early blight test means the compound is not a fungicide, and terminal branching of a cotton plant in a boll weevil test means the compound is not an insecticide. In each case, the net result of a stereotyped screening program is the placing of compounds on the shelf where they become a part of the tremendous inventory of potentially useful, but dormant, compounds.

Fortunately, growth-regulant screening has come a long way. Most major companies now recognize this effort as a significant part of their program. Activities are directed toward method development whereby desirable responses can be detected and correlated to commercial needs at an early stage. Follow-up is conducted on a routine basis, and field testing is accomplished as dictated by the compound's activity. As a result, it is reasonable to project a tremendous increase in the availability of growth regulants for research and development purposes.

Today's commercial growth regulants originated at a time when they suffered the difficulties, as well as enjoyed the benefits, of being the first to be discovered. Perhaps the greatest difficulty with some of the first growth regulators was the fact that the responses observed in screening programs did not seem to have the broad application potential of other pesticidal products. It really wasn't until horticultural researchers showed the potential value of these materials in such specialty areas as floriculture, vegetables, and pomology that there was a significant shift in interest to pursue the commercialization of these exciting new compounds. In retrospect, it is now obvious that the successful development of growth regulators was keyed to the premium the horticultural industry was willing to pay for products which would reduce labor costs or increase crop values.

The production of horticultural crops is based upon programs which manipulate virtually every part of the plant's environment. It is not uncommon for the grower to deliberately modify the quality and quantity of light, daily and seasonal temperatures, quality of the air, type and fertility of the support media, moisture content of the soil and air, and the size and shape of the plant. As a result of these sophisticated cultural programs, it is possible to produce fruit and vegetable crops with values exceeding several thousands of dollars per acre and florist crops with values in excess of a hundred thousand dollars per acre.

Yield obviously is a major consideration in the production of any crop, but the additional value associated with a different grade, a different harvest time, an improved storage quality, or an ability to use mechanization in the growth and harvesting programs is substantially greater in horticultural than in agronomic crops. The ability of the horticultural industry to pay a price which encouraged the development of compounds for specialty markets provided the incentive for chemical companies to break with their traditional desires to find products with multimillion-pound potentials. This is not to say that the interest in large-volume growth regulants has been diminished but rather that profitable markets have been found in specialty crops.

The apple industry offers several good examples of specialized cultural program requirements which encouraged the commercial development of growth regulators. Apple growers have historically tried to modify tree form and size. Until recently, the basic tools to accomplish this included the selection of root and scion stocks and the amount of time and labor available to prune the trees mechanically. Striking the balance between the trees' inherent capability to grow and the farmers' ability to manipulate the tree with his pruning shears was a constant battle. The development of the growth regulator daminozide provided the fruit grower with a chemical tool that not only suppressed the rapid flush of undesired growth which often follows mechanical pruning but also encouraged the development of flowers and fruits rather than stems and leaves.

More recently, it has been demonstrated that an apple tree's reproductive capability can be further regulated by the combination of chemical thinning programs and growth retardants. The ability of such programs to overcome biennial bearing problems has a specific economic value. For apple cultivars which tend toward biennial bearing, it would not be unusual for a potential 500-box orchard to produce only 200 boxes in the off year. Appropriate programs with growth regulators that combine thinning and retardant activities have been able to prevent the off years and maintain production at optimum levels. An important point which must be understood is that the $2000 to $3000 additional value of the crop is based upon the effective use of growth regulators in a program rather than on the use of the chemicals themselves.

In a similar manner it is not possible to determine the precise economic contribution of daminozide's growth-retarding activity in a meadow-orchard concept of producing apples. In this experimental system of producing apples, which has been pioneered by the Long Ashton Research Station in England, apples are planted at a density of 70,000 per hectare (approximately 28,000 per acre) and treated with daminozide to initiate flowers in their first year of growth. During their second year, the trees flower and produce fruit, after which they are cut back to a stump from which a new shoot is regenerated to repeat the biennial cycle. This system is capable of producing yields of up to 800 bushels per acre, but it is essentially designed to facilitate full mechanization of harvesting and pruning. The key point which must be understood about the meadow-orchard concept is that it is a program which fully integrates the biological, mechanical, and chemical tools available to the fruit grower.

An interesting new chemical tool which will probably find its way into future fruit production programs is the chemical pruning agent. Tree shaping and form maintenance have conventionally been the direct result of mechanical pruning. Chemical growth retardants have been useful in manipulating the resultant growth from mechanical pruning, but the ability to prune chemically is just coming into the picture.

Historically, it has been observed that chemicals which destroy or inhibit the apical meristem also induce abnormal axillary growth. What are needed are compounds capable of stopping terminal growth and stimulating vigorous, uniform side shoots. The esters of the fatty alcohols, which were originally found to be tobacco-sucker regrowth-control chemicals, have been used to kill selectively apical meristematic tissue and induce branching on many ornamental plants. Research is continuing in the fruit areas. Uniroyal has recently discovered a unique compound which induces the so-called feather type of growth on apples that is normally obtained only through a dedicated and detailed mechanical pruning program.

This same compound can be used to retard the development of both vegetative and reproductive axillary shoots if it is applied at a time when the axillary meristems are preferentially more susceptible than the terminal meristem. Work is currently in progress to evaluate this compound as a partial substitute for hand removal of undesired lateral flower buds from chrysanthemums. Estimates have been made that the hand-labor costs to remove floral disbuds from a 100-foot bed of standard mums is at least $24–40. It is expected that the chemical can be used in such a way that at least 50% of the disbudding operation can be accomplished chemically at a substantial savings in labor cost to the florist.

Another interesting potential application for this type of material would be in the production of greenhouse tomatoes. The 1971 Annual Report of the Glasshouse Crops Research Institute from the United Kingdom estimated that approximately two billion side shoots were removed annually from greenhouse grown tomatoes in the U.K. at a cost of about 500 English pounds or $1,250 per acre. This was equated to be 5% of the total production costs. Obviously, this is an area which closely corresponds to the concept of limited market needs which can afford to support the price of developing new compounds.

Many other examples exist where vegetative growth regulating activity has been demonstrated to be of economic importance to horticultural crops. The research efforts of the past decade have provided the florist with at least a half-dozen different chemicals with which he can retard the vegetative growth of virtually any crop he wishes to grow. Potted chrysanthemums serve as a good example where the need to stake and tie the plant to prevent it from falling apart has been virtually eliminated by the use of vegetative growth retardants. Tall, leggy lilies, hydrangeas, poinsettias, shaggy azaleas, and poorly branched bedding plants are now the exception rather than the rule.

In the case of vegetable crops, where the consumer could care less about the appearance of the plant producing the crop, vegetative growth retardation takes on a different value. Mechanical harvesting of many vegetables is a question of separating the desired crop from the plant. The problem is complicated by the fact that mechanical harvesting of vegetables is usually a once-over procedure. This means that the uniformity of the crop at the time of harvest becomes a critical issue. Modifications in vegetative growth which improve the harvester's ability to separate the crop from the plant can be accomplished by compounds which either reduce overall plant growth or selectively kill or remove undesired vegetative growth from the plant.

During the past few years, a concerted effort has been made to develop a machine which will pick tobacco leaves. Prototypes quickly evolved into commercial machines which during the course of a single day can pick eight acres of tobacco. A conventional hand-picking operation would require six people walking the same area to do the same job. Despite the fact that a machine will cost $16,000 and have a life expectancy of 10 years, it is now projected that any farmer with more than 30–40 acres of tobacco will harvest mechanically because of the price and more importantly, the lack of farm labor.

The relationship between the development of the tobacco harvester and chemical growth regulators is one of direct interaction. One of the major problems that needed resolution was the inability of the mechanical harvester to distinguish a leaf from axillary regrowths. If these axillary regrowths, which are more commonly known as suckers, are allowed to grow, they can so thoroughly confuse the mechanical harvester that it is virtually impossible to operate.

Fortunately, the tobacco farmer has two types of growth regulants which allow him to produce sucker-free stalks. The first material acts as a local contact material and prevents the development of axillary suckers at a time when the leaves are developing their maximum size and weight. The second material has a systemic action and is applied approximately 10–14 days after the contact treatment. It moves throughout the plant to hold sucker growth during the actual harvest period. As a result, it is possible to produce a plant with a clean stalk, and there is a net increase of about 200 pounds in the weight of the harvested leaves. In terms of dollars, this means an additional $160–250 in yield plus the flexibility of labor management through the use of the mechanical harvester.

The story is far from complete in tobacco because new growth regulants are now being evaluated which will cause all the leaves on the plant to mature at one time. Conceivably the day will soon be reached when a program based upon chemical growth regulants and mechanization will reduce the harvesting of flue-cured tobacco from five trips to one.

Concentrated maturity and associated improvements in quality is the other portion of the harvest picture which has so greatly benefited from plant growth regulators. Ethephon has been used to increase the red coloration of apples with a net improvement in grade. For every box of apples which is graded up to fancy from #1s, there is a premium of $2–4. It is not uncommon for a $25 ethephon treatment to shift 50% of the production up to fancies, which translates out to $500 of additional value on a 500-box yield. There are other crops where both daminozide and ethephon have been found to be effective in concentrating maturity and aiding mechanical harvesting techniques. These include tomatoes, cantaloupes, peaches, and cherries to mention a few.

Growth regulants are not only capable of influencing the quality of the crop at harvest time, they can also dramatically alter the harvest time itself. The gibberellins, daminozide, and ethephon are well known for their respective abilities to hasten or to delay the maturity of such crops as cherries and peaches. Even though it is well understood that every grower can not be first to the market, the 2–10 × premium associated with bringing in the first fruit makes the cost of growth regulators an insignificant part of the program.

From the standpoint of efficient harvesting techniques, the hastening or retarding of maturity plays an important role in determining the amount of capital a grower must invest in harvesting equipment. As good as a $15,000 mechanical tart cherry picker may be, it can only pick so many cherries in a day. By the proper use of growth regulants, it is possible to extend the time a single harvester can operate by an additional three or more weeks. The benefits which accrue to the tart cherry grower in the form of reduced capital and labor costs are further supplemented by the greater flexibility offered the processor, who can run his operation at a more constant pace over a longer period of time.

A classic example of spreading the harvest is found in the citrus industry where growth regulators are used to reduce the acidity of grapefruits. This practice allows the harvest of grapefruits for almost a nine-month period. There is no way to put an economic value on this particular practice. It involves too many people, too much equipment, and too many dollars to quantify the benefits.

A more easily defined situation can be seen in production of Mac-Intosh apples. These apples, which make up a very significant portion of the northeastern U.S. apple crop, are notorious for dropping off the tree as soon as they have reached maturity. Prior to the development of stop-drop chemicals, it was expected that at least 20–25% of the crop would go on the ground before harvest. The stop-drop programs of today can virtually eliminate this problem, spread the harvest over a longer period of time, and improve the quality of the fruit as it is sold fresh or as it comes out of long-term storages. The value of the stop-drop effect alone could be calculated to be at least $500 on a 300-box-per-acre yield.

No discussion of horticultural growth regulators would be complete without some comments about abscission agents. In this area, the efforts of the Lake Alfred citrus research group in Florida to find citrus loosening agents and to integrate them into mechanical harvesting programs are particularly noteworthy. It was determined at an early date that even though the bulk of Florida's orange production was oriented to the juice market, there was a very distinct need for a citrus fruit-loosening agent which would assist the mechanical harvesters. Cycloheximide was one of the products developed as an answer to this problem.

Until the recent recession, the economic value of cycloheximide was usually associated with the development of air-harvesting equipment. In 1973 a portion of the Florida crop was never picked because hand labor was unavailable. The forecasts for 1974 and 1975 indicated that mechanical harvesting was a must if the industry was to survive. The recession hit, and farm labor became more available. Instead of decreasing in importance, cycloheximide actually grew because the efficiency of hand-labor picking was significantly improved. As a result, the importance of fruit loosening took on another dimension for the citrus industry.

The final, and most obvious, economic value of growth regulators is their ability to increase crop yields directly by inducing more harvested fruits or pounds per acre. When compared with some of the relatively sophisticated uses within the framework of special cultural programs, it may sound rather mundane to mention the 15–25% increases in yield of

Concord grapes, peas, seed alfalfa, and seed clover as a result of damin-oxide treatments or the 50% increases in yield of pickling cucumbers because of ethephon. In addition, there are many crops where 10–15% increases have been consistently noted, but because of normal crop variability, these increases are not considered to be of economic importance. It would seem that it is only a matter of time before more active compounds are found or systems devised whereby available products can be utilized to provide the desired yield responses.

It appears that the economic value of growth regulators to producers of horticultural crops is exceedingly high. The normal criterion of returning $3–4 for each dollar invested is often exceeded many times over in direct returns to the grower. If the indirect benefits are considered, then it is no small wonder why these products are enjoying continued growth and commercial success. Fortunately, the horticultural industry has both the need as well as the ability to pay the price necessary to support the development of products for relatively small, specialized uses. The benefits which will accrue from the products and knowledge being developed in these areas will most certainly lead us to the development of growth regulators in all areas of crop production.

RECEIVED February 22, 1977.

8

Plant Growth Regulator Potential on Sugarbeets

E. F. SULLIVAN

Agricultural Research Center, The Great Western Sugar Co.,
Longmont, Colo. 80501

Plant growth regulators used on sugarbeets should improve crop emergence rate, enhance production, and conserve sugar produced. Experimental results indicate that foliar applications at canopy closure are more likely to improve yield, while applications three to six weeks pre-harvest are more effective for promoting sugar content. Plant population and spacing affect sugar content, and excessive soil nitrogen causes low sugar and high root impurities at harvest. Three maturation points occur during later growth— namely, nitrogen depletion aging, temperature senescence, and finally, low temperature-induced growth cessation. Yield improvement by chemically promoting growth before August seems possible. Chemical regulation of root/top ratio later and adjustment of senescence to promote greater and earlier sugar buildup in the root are reasonable aims.

Useful plant growth regulator candidates on sugarbeets should accomplish one or more of the following objectives or benefits: (1) improve seedling emergence rate and vigor to permit planting to final stand without supplemental adjustment; (2) increase root yield; (3) enhance root quality; and (4) conserve sugar produced during storage after harvest by chemical regulation of root respiration rate. The average emergence rate is 55% from a monogerm seed with a germination potential of 95%. Generally, sugarbeets are overplanted to offset stand loss caused by an adverse environment. Subsequently, the stand is mechanically or manually adjusted to the final or harvest stand. Supplemental stand adjustment is a costly practice.

Essentially, experimental growth regulators are applied on sugarbeets to augment or impede natural growth processes and rates—e.g., to
quicken seed germination, crop emergence, and seedling growth early
and to retard top growth later on. Effective chemical systems are expected to quicken leaf canopy closure time by 10–14 days, which expands
the growing season and yield. Sugarbeets have a relatively slow growth
rate from soil emergence until the crop leaves cover the row. A better
understanding of industrial methods, limits, and objectives might well
serve both the agricultural chemical supplier and researcher in advancing
their search for an effective and usable plant growth regulator on
sugarbeets.

Grower Considerations

Growers receive direct benefit from producing the highest sugar
yields possible per acre. High sugar producers are capable of 10–12,000
lb per acre. This accomplishment requires expert management because
root weight and sugar percentage are agronomically inversely related
generally. Regulating the two factors beneficially and simultaneously
from chemical application has not yet been demonstrated, although single
factor adjustment has been attained. For example, gibberellic acid (GA_3)
and 2-(chloroethyl)phosphonic acid (Ethrel) improve root yields but
decrease sugar concentrations, whereas maleic hydrazide (MH) produces sugar increases which are inversely related to root weights.

Today, field labor costs have little or no relationship to pricing crop
protection and production chemicals. Current production is based on
crop chemical technology rather than on labor use. Growers will sched-
ule costs for improving uniform crop emergence rate and sugar yield
per acre because of direct participation in benefits. The prices that
growers and processors receive for sugar sets a practical upper limit on
plant growth regulator cost and use. This is so because, as with all
agricultural chemicals, the benefits derived from plant growth regulators
must exceed the costs of application.

Growth Environment, Yield, and Sugar Content

As in other crops, response reliability among years and sites remains
a major problem when sugarbeets are treated with plant growth regulators. Application timing, variety selection, soil nitrogen fertility level,
moisture and temperature regime, and other variables more than likely
interact and regulate response magnitude. Nevertheless, experimental
results indicate that topical applications made early in the growing
season (from the 12-leaf stage until closing of the rows) are more likely
to improve root yield, while applications made three to six weeks before

harvest are more effective for promoting sugar content. Plant growth regulator mixtures and split or sequence applications have been generally ineffective. Sugarcane ripeners, namely, N,N-bis(phosphonomethyl)glycine (Polaris), 3-trifluoromethylsulfonamido–p-acetotoluide (Sustar), and methyl–3,6-dichloro–o-anisate (Racuza), have been ineffective when applied topically on sugarbeets in trials conducted by The Great Western Sugar Co.

In the irrigated inter-mountain and the Great Plains regions, sugarbeets require approximately 180 days from planting to harvest for maximum sugar production. Post-thinning stands of 24–28,000 plants per acre usually diminish somewhat by harvest. After a week or so of recovery from mechanical thinning, seedling beets begin growing more rapidly since root systems are well developed, especially after row closure. Results from several proprietary compounds reveal that topically applied growth regulators for root weight stimulation should be applied when seedling beets have about 12–14 true leaves or when the canopy is about 80% closed. Approximately 120 days are required to obtain the full effect from a treatment. Effective regulators applied at row closure are expected to increase the root weight by 1.5 ton/acre and to improve sugar yield by 6–8%. In 1975, UC-51416 (confidential chemical structure) gave preliminary promise as a sugarbeet yield-enhancing chemical (> 3 tons root weight per acre). Additional screening of new compounds and more rapid and reliable screening methods are needed to advance the use of plant growth regulators on sugarbeets. To date, no plant growth-modifying chemical has been released for commercial use on sugarbeets.

Although plant population and spacing affect sugar content, an excessive soil nitrogen fertility level is the major cause of low sugar and high root impurities at harvest. A sidedressing of nitrogen in beet fields is discouraged after July 15 to ensure soil nitrogen depletion and plant uptake at a declining rate. Most beet sugar companies have effective nitrogen fertility monitoring programs based on a production ratio of 8–10 lb of available nitrogen per ton of beets.

In practice, three maturation points occur in the late growth pattern. Nitrogen depletion aging (physiological senescence) is initiated between August 15 and September 1 (six weeks before harvest). At this point petiole nitrate nitrogen levels should decline to 1000–1500 ppm for optimal sugar content at harvest. It is expected that a timely plant growth regulator application at this point will further suppress top growth and nitrogen accumulation (amino nitrogen) in the plant, especially if soil nitrogen is above critical levels, which usually occurs. Growth rate of the tops is naturally declining at this point and sugar buildup commences more rapidly. Growth regulators applied at temperature senescence or at the point where night temperatures significantly decline (first frost) and daytime growing conditions are warm and sunny may also be effective for sugar increase. Plant growth modifying chemicals are expected to lower plant nitrate levels which at high levels depress sugar content of roots. Effective chemicals applied at the late growth period should increase the sugar content of roots by 0.5–0.75%. Harvest roots usually contain 15–17.5% sugar. True senescence or plant growth cessation occurs on or about November 1 or when temperatures reach 26°F or lower for 6–8 hr. When plant growth stops, sugar accumulation stops.

Quality Considerations

Substantial savings can be realized by root purity improvement from agronomic means such as variety improvement or from an effective chemical application. Harvest impurities consist primarily of amino nitrogen, betaine, and potassium, sodium, and chlorine.

Direct delivery beets (sliced soon after delivery without piling) have the greatest amount of extractable sugar. Beet piles are usually covered with straw or plastic sheeting to protect against deep rim freeze. Storage impurities and root decay organisms which invade harvest wounds increase as the storage period lengthens. Sucrose losses throughout storage are estimated at an average of 0.5 lb refined sugar per day per ton of beets. Main storage impurities from prolonged root respiration periods causing sucrose dilution are non-sucrose sugars—namely, glucose, fructose, and raffinose. A plant growth regulator that conserves sucrose during storage by lowering respiration rate is a worthwhile objective but of more practical value if linked with field quality improvement.

Risk-Benefit Evaluation

Yield improvement beyond genetic and management inputs by chemically promoting top growth before August seems possible. Chemical regulation of root/tops ratio later in August and adjustment of senescence to encourage translocation of sugar to the root for earlier harvest are worthwhile objectives. A plant growth regulator that improves sugar yield seven years out of 10 would have sufficient field efficacy for chemical development and grower use. The risk-benefit margin may limit future cost of the chemical per acre to the average production cost per ton of beets since profitable farm production goals are primarily based on crop protection chemicals, nitrogen fertility, and adapted varieties.

RECEIVED September 21, 1976.

9

Economic Potential of Growth Regulators for Floriculture and Woody Ornamentals

A. E. EINERT

Department of Horticulture and Forestry, University of Arkansas, Fayetteville, Ark. 72701

Two classs of growth regulators, root promoting compounds and retardants, dominate commercial floriculture and woody ornamentals production. These seem to offer the most future economic potential. The market for rooting compounds will probably continue to grow because of new, hard-to-propagate cultivars, increased production, and efforts toward increasing production efficiency. Commercial synthetic retardants are now supplementing the formerly exclusive realm of IAA, IBA, and NAA for this purpose. Retardants afford the most potential for ornamentals. The growers' collection of compounds is being replaced by new products providing options of rate, application method, and timing. Retardants commanding the future market must be effective on a wide range of crops, provide versatility of application in many cultural systems, and retard senescence in the harvested product.

In 1968 a joint effort by the USDA and state agricultural experiment stations prepared a document entitled "A National Program of Research for Plants to Enhance Man's Environment" (1). This task force addressed itself to research questions needing answers in floriculture and ornamental horticulture. The report listed two topics for plant growth regulators (PGRs) in the production of ornamentals and only two references to the future role of PGRs in the establishment and maintenance of landscape plants. Specific requests were made for more efficient and effective methods of propagation, suggesting a need for PGRs and for investigation into the nature and activity of native plant growth substances. The report also called for study of synthetic regulators on morphogenesis, especially with regard to dormancy, apical dominance, height control, and flowering. A special request was made for agricultural research to develop retardants which would reduce turf mowing and plant maintenance costs.

This task force report was prepared eight years ago. During the intervening years, PGRs have been woven into more far-reaching areas of ornamental plant investigations. A Southern regional group formed recently to update research priorities discussed such additional possible future roles of PGRs as: adapting native plants (and foreign introductions) to cultivation; adapting plants to mechanized production systems; adapting plants to the more unfavorable growing environments of urban areas; maintaining quality during distribution; and adapting plants to future energy conserving production environments. The concept of adapting or altering the plants to the changing environment of production and use implies a shift in emphasis toward needs of people rather than plants. Obviously, the ideal situation will be modification of both the plant and the cultural system.

Before attempting to assess future roles and the economic impact PGRs will have on commercial floriculture, woody ornamentals, and turf, it is well to determine the present scope of their use. Rather than enumerate PGR compounds and their particular niche in production schemes, the plant growth phase being influenced will be highlighted. This is especially warranted since the traditional role of certain classes of compounds, such as indole acids in vegetative propagation, is now being formidably challenged by other compounds—namely ethylene and antigibberellin retardants.

The prime areas of growth regulation for ornamentals, discounting for this discussion the herbicides, are: (a) propagation, still dominated by growth promoters, (b) height control, presently in the realm of retardants, yet now including chemical pruning agents and morphactins, (c) regulation of flowering with limited PGR use of promoters and retardants, and (d) extension of post-harvest life involving many new combination products such as vase-life prolonging formulations.

The four categories of plant responses will undoubtedly show increased PGR use in the future, yet it appears that propagation and plant height control may afford the most ready outlet for new commercial PGR products. Senescence retardation and morphological manipulation present real challenges for basic exploratory and developmental research and promise new product potential in the distant future. Economic potential will be recognized easily with imaginative use, even with the compounds presently available.

Plant Propagation

A vast number of ornamental plants in the commercial market are asexually propagated by cuttings and vegetative structures such as bulbs and corms. Virtually all woody plant cuttings are treated with root promoting compounds containing active ingredients of IAA, IBA, NAA, and perhaps boron. Consumption of these products will increase in direct proportion to increases in numbers of plants propagated. Other compounds that have recently been shown experimentally to enhance propagation are ethephon, which Sanderson at Auburn University is evaluating on various woody species, and the retardants, SADH and chlormequat, which workers in Idaho (2) have reported are beneficial on succulents.

In addition to products already marketed, research on extraction of additional natural root-promoting materials (3) will undoubtedly offer future potential for commercial synthesis. Gibberellins, cytokinins, and morphactins are also interesting possibilities for more precise regulation of seed propagation. For bulb crops, Nightingale at the Texas Agricultural Experiment Station has promoted bulbing in lilies with the cytokinin, SD–8339.

The largest expansion in PGR use in propagation will probably result from increased commercial production. The increasing interest in plant production and reproduction by the general public, however, should further expand the market. One of the prime factors for increased interest in propagation (commercial and private) will be the introduction of new plants into cultivation or the adapting of native plants to large scale production schemes. Newer cultivars which are difficult to transplant or root may enter the commercial production arena (4) by new application techniques of standard rooting compounds and new products, probably containing ethylene precursors and morphactins.

Plant Height Control

Height control has been the most economically fruitful area of applied PGR use and still suggests more potential. Almost every commercial greenhouse-grown floricultural crop relies on a growth retardant in its production. Before the advent of the host of commercial retardants introduced in the past 20 years, all sorts of height-limiting techniques were used, such as physically tying the stem back on poinsettia, withholding water, and reducing daylength. Growth retardants opened up cultural possibilities by providing all kinds of ways to manipulate plant height. Not only have they permitted growers to meet size requirements set by the consumer, but they also have established new product types, for example, the mini-poinsettias and other small plants for which few genetic dwarfs presently exist.

Although retardants have been a boon to floriculture, new developments are needed. A commercial grower now must stock a variety of retardants since the "accepted" PGRs differ among the major floriculture crops. SADH (B–Nine) is the commercial standard for chrysanthemums and bedding plants, CBBP (Phosfon) for lilies, and chlormequat (Cycocel) for poinsettia. This situation suggests a regulator specificity when in fact grower acceptance is based on satisfactory previous experience. Formulations which are easy to apply and low in phytoxicity make certain compounds desirable to growers. Many growers are not presently aware, for instance, that the recently marketed ancymidol (A–Rest) is effective on every crop listed above and offers many options to fit the growing operation with respect to rate and number of applications as well as method of application. This regulator can also influence the postproduction plant performance.

The ideal retardant for future commercial acceptance will be a universal product effective on all crops, offering application alternatives; it should be non-phytotoxic and will enhance post-harvest life without any harmful residues under commercial conditions. Research at Arkansas is heavily directed toward investigating alternate modes of applying PGRs and determining the fate of residues (5, 6). In fact, attempts are being made to recycle residues in second crops.

The direct material cost for a growth retardant, based on updated calculations of Griffith and Payne (7), is about $0.04 per 100 pots of chrysanthemum under efficient operating conditions. Labor to apply the retardant (at present wages) is an additional $0.25, yielding a total cost of $0.29 per 100 pots. With chrysanthemums, the retardant is a single spray of B-Nine; this cost would be higher with drench application or a more costly retardant. A 1975 production and marketing survey of flower growers in Philadelphia, Baltimore, and Washington, D.C. (8) listed the value of potted plants from 105 firms at $3.5 million. The value of the growth retardant used for this production level (based on rough estimates of use and price) would be approximately $1,050. Flower production increased nationally about 200% in the 10-year period between 1963–1973. Foliage in the three-city survey area increased 49% between 1966 and 1971 and jumped by 76% from 1971 to 1973. Foliage plant production is booming nationwide and is expected to maintain its growth somewhere between these two percentage levels.

Another area of strong potential for retardants is landscape plant maintenance. The commercial woody plant producer needs to accelerate growth and increase plant height and size as rapidly and efficiently as possible. Most woody landscape plants are judged for price and quality by height. These same plants in the landscape, however, are required to grow to a certain maximum height and form dictated by a specific design and to maintain this form and space relationship with a minimum of maintenance. A prime example of this is hedge plantings. Although many dwarf and slow-growing cultivars are available, the demand is for a plant which grows rapidly to a desired landscape height. Growth regulators can be used to initiate and terminate growth when desired. Most current research efforts are directed primarily toward compounds which terminate growth. Growth promoters such as gibberellins and cytokinins could be incorporated into future plant-establishment practices to initiate early growth.

Research is needed to devise practical systems of pruning and PGR control, especially since outdoor environments can drastically alter activity. In terms of economic potential for such products, a study on the University of Arkansas campus (9) documented a 37% saving of labor required to prune hedges of several plant species over a two-year period using sprays of an experimental retardant (NIA 10637) during the first year of the experiment. In addition to reduced labor cost, the retardant caused a deeper green foliage color, allowed a delay in the initial spring pruning the second year, and discouraged insect pests.

A consideration of growth control in landscape plants must also include PGRs formulated as chemical pruning agents, sprout inhibitors, and the expanding possible use of morphactins. PGRs—specifically gibberellins, cytokinins, and ethylene—have been used to alter growth habits by reducing apical dominance and prolonging juvenility, thereby completely changing growth characteristics. This could open up new markets for a particular plant.

The practical application of retardants and promoters to turf is also receiving concentrated research attention and offers future promise of expanded commercial markets. State experiment station scientists are stimulating growth of southern grass species during cool weather with gibberellin applications (Dudeck–Fla.), delaying spring growth with retardants (Burns–Ga.), and antagonizing seed-head production in grass with retardants (Ward, Coats, and Laiche–Miss.).

Regulation of Flowering

Many floricultural crops are in peak market demand only during certain periods because of their public association with particular holidays. Crop blooming is presently timed commercially by manipulation of temperature and light, thereby altering endogenous growth (and flowering) regulators. Exogenously applied PGRs can trigger, enhance, hasten, or delay the onset of blooming in a few crops. If research can elucidate the factors controlling flowering, new and economic possibilities for PGRs will be developed.

Extension of Post-Harvest Life

Post-harvest and post-production behavior of ornamental plants have brought forth a new class of PGRs commercially called vase solutions. Research with these materials, largely by Marousky (*10*), now enables the extension of cut-flower life and allows flowering stems to be cut in the tight-bud stage and develop normally during storage and transit. Vast potential lies here for PGR compounds which retard senescence to be used by retail florists and even the homeowner. Study of transportation problems with ornamentals has seen emphasis on the control of native growth regulators, ethylene and CO_2, through hypobaric storage (*11*). Commercial retardants with long-lasting activity, applied during the growing period, offer future possibilities for extending flower life in the consumer's home. Stem weaknesses have been reduced experimentally in cut tulips by pre-harvest sprays and vase solutions of ancymidol (*12, 13*). Practical application of such findings for major cut-flower crops is desperately needed by the retail flower industry.

Summary

The overall potential of PGRs for ornamentals seems to exceed that for all other commodities. This optimistic forecast is the result of the expected increase in production of these crops and the fact that ornamentals are non-food crops and have not been associated publicly with such terms as chemicals, pesticides, and herbicides, with all their popular negative connotations. Growth regulator manufacturers in the future must market complete, integrated management systems using PGRs and not merely sell bottles of miracle compounds.

Literature Cited

1. USDA and State Universities and Land Grant Colleges, "A National Program of Research for Plants to Enhance Man's Environment," Unnumbered publ. of Joint Task Force, *Res. Prog. Develop. Eval. USDA*, Wash., D.C. (1969) 30 pp.
2. Boe, A. A., Stewart, R. B., Banko, T. J., "Effects of Growth Regulators on Root and Shoot Development on Sedum Leaf Cuttings," *HortScience* (1972) **7** (4), 404.
3. Heuser, C. W., Hess, C. E., "Isolation of Three Lipid Root-Initiating Substances from Juvenile *Hedera Helix* Shoot Tissue," *J. Am. Soc. Hortic. Sci.* (1972) **97** (5), 571.
4. Einert, A. E., "Propagation of Dwarf Crapemyrtles," *Proc. Int. Plant Prop. Soc.* (1975) **24**, 370.
5. Einert, A. E., "Slow Release Ancymidol for Poinsettia by Retardant Impregnation of Clay Pots," *HortScience* (1975) **11** (4) 374.
6. Williamson, C. L., "Ancymidol as a Granular Slow Release Retardant for Potted Poinsettia," M.S. Thesis, University of Arkansas (1976) 30 pps.

7. Griffith, H. V., Payne, R. N., "An Analysis of Pot Chrysanthemum Production Methods, Direct Costs and Space Use," *Okla. State Univ. Agri. Res. Bul.* (1969) B-670, 39 pps.
8. Hall, R., Raleigh, S., "Marketing Practices of Growers of Flowers and Plants: Philadelphia, Baltimore, Washington, D.C.," *USDA, ARS Bull.* (1975) 593, 42 pps.
9. Einert, A. E., "Evaluating Chemical Pruning for Hedges," *Ground Maint.* (1973) **8** (1), 24, 25, 56, 57.
10. Marousky, F. J., "Recent Advances in Opening Bud-Cut Chrysanthemum Flowers," *HortScience* (1973) **8** (3), 199.
11. Burg, S. P., "Hypobaric Storage of Cut Flowers," *HortScience* (1973) **8** (3), 202.
12. Einert, A. E., "Reduction in Last Internode Elongation of Cut Tulips by Growth Retardants," *HortScience* (1971) **6** (5), 459.
13. Einert, A. E., "Effects of Ancymidol on Vase Behavior of Cut Tulips," *Acta Hortic.* (1975) **41**, 97.

RECEIVED August 27, 1976. Published with the approval of the director of the Arkansas Agricultural Experiment Station. Use of a trade name does not imply endorsement or guarantee of the product or the exclusion of other products of similar nature.

10

The Development of DNBP (Dinoseb) as a Biostimulant for Corn, *Zea Mays* L.

A. J. OHLROGGE

Department of Agronomy, Purdue University, W. Lafayette, Ind. 47907

Seven years of field experimentation demonstrated a high probability of increasing corn grain yields 5–10% with low rates—3–6g/a (acre)—of 2-sec-butyl-4,6-dinitrophenol. The DNBP is broadcast over the foliage when the unemerged tassel is between 0.5 and ca. 7 in. long. Yield increases are usually associated with about two days' earlier pollination, less barrenness, and greater weight per ear. Physiological maturity appears to be unaffected by treatment and therefore the longer filling period contributed to greater yield. Relationships in the community of plant growth regulator (PGR) developers are briefly discussed.

In 1968 T. G. Sherbeck and E. S. Oplinger (two of my graduate students) screened 17 PGR chemicals on corn. Many of these chemicals were herbicides and were added to the starter fertilizer band applied on corn. Why in the fertilizer band? Earlier work with TIBA (3,5-triiodobenzoic acid) in the fertilizer band on soybeans had shown promise for increasing grain yields (1).

One chemical, DNBP, stimulated earlier vegetative growth which, in turn, prompted us to apply the chemical on the corn foliage during that same year. At this point I was not aware of the stimulatory effects of the dinitro compounds as reported by Crafts (2) in the early 1940s or more recently in the experiments of Bruinsma (3) in the Netherlands. In his 16 experiments conducted between 1953 and 1962, DNOC, 4,6-dinitro-o-cresol, gave increases in winter rye ranging from 0 to 28%. The mean increase was 6%.

The effects of DNBP in the fertilizer band are shown in Table I. Both plant height and grain yields were positively affected by the small quantities of DNBP in the fertilizers. No further work has been done in

79

Table I.   The Influence of DNBP in a Phosphatic Fertilizer Band on the Growth and Grain Yield of Corn: Purdue Agronomy Farm (1968)

DNBP (g/ha)	Plant Height (cm)	Grain Yield (kg/ha)
0.0	136	8880
2.5	149	9190
12.5	147	9400
62.5	144	9270
312.5	142	9640

this area because of the great complexities of the root–soil–fertilizer band system and because of the rewarding returns from foliar applications of DNBP.

Early Experiments

In 1969 D. Hatley (4) established rate × time factorial experiments on three cooperating farmers' fields. At Gaston, Ind., corn grain yield increases and the earlier pollination times were statistically significant at the 5% level. The hybrid grown was XL45. Because of uneven tasseling, uneven emergence of the seedling corn, and the fairly low yield level, the Coatsville site was believed to be a poor test of DNBP. The third site at Monan, Ind. gave excellent yields but little positive response. This field had been treated with 2,4-D, and the corn was brittle at the time of DNBP application. Whether or not this was a complicating factor is not known. In the use of PGRs, interactions with herbicides must be taken into consideration. In addition to Hatleys' experiments, Winter (5) established an Ethrel[(2-chloroethyl)phosphonic acid] × Premerge (2-*sec*-butyl-4,6-dinitrophenol) experiment at the Purdue Agronomy Farm. Statistically significant (10%) increases in grain yields were obtained when DNBP as Premerge was used alone.

Financial Concerns

How was the work of 1968 and 1969 financed? A major PGR development program for TIBA on soybeans had been financed at Purdue largely by the International Minerals and Chemical Corp. Residual money from these grants-in-aid helped support the DNBP work. The increased yields at Gaston and the Agronomy Farm suggested that we might have the beginning of a breakthrough in the use of PGRs on corn. With this conviction firmly in mind and with the support of the Purdue Research Foundation we contacted on a confidential basis about five of the major agricultural chemical companies. We hoped to secure patents and development money. Our efforts were futile: not one dollar of support was secured. In general the answers were the same: the resarch results looked most encouraging, but since the manufacturing patents had expired, protection would depend on a use patent, which is almost useless in practice. In retrospect, another important facet in the corporate decisions must have been the extremely low rate of usage of the chemical. At current retail prices it was about two cents per acre, which, under the most favorable conditions, could not add up to a very large dollar volume of business.

While we were seeking commercial support, one replicated experiment per year was conducted at the Purdue Agronomy Farm. Results continued to show promise although the optimum rate seemed to get higher as each year passed. The DNBP source provided an explanation. The same bottle of reagent which Hatley had obtained from the herbicide research group at Purdue in 1968 was used in 1972. Thus, it was at least five years old and may well have been six or seven years old. Shelf life according to the herbicidal handbook is two years (6).

Decisions to be Made

By the winter of 1972–73 I had given up the hope of getting financial aid from industry to study the response of corn to DNBP as a grain yield enhancer. Should we place the DNBP research notebooks on the shelf and go on to other things? Before I took this final step we decided to look at the data again. What could a grower expect from the application of the optimum rate of DNBP? Because of the aging of our DNBP source, Dow Chemical Co.'s Premerge, it was difficult to determine the optimum rate. We decided to look at the best and the poorest responses relative to the untreated corn in each of the experiments. If the increases were caused by random variations, then the mean of the poorest responses should equal the mean of the best responses. The results are shown in Table II. The mean increase of 13 bu/a for the best response and 0 bu/a for the poorest responses renewed our enthusiasm, especially since the

Table II. Best and Poorest Responses of Corn to DNBP in Rate Experiments

Year	Poorest (bu/a)	Best (bu/a)
1968	−6	16
1969	+5	12
1969	−1	13
1970	+2	19
1971	−1	16
1972	+1	7
Average	0	13

13 bu of corn grain were obtained for a chemical cost of only two or three cents per acre. Equally important were the indications that rate of application was not extremely critical. This was quite different from our experiences with TIBA where even the time of day influenced its effectiveness (7).

The economics of our data indicated that a farmer would probably obtain a return of about five to 10 dollars for each dollar he invested in the application of DNBP on the foliage of corn. Such handsome returns not only excited me but also farmers who listened to our reports at Agronomy Field Day or read stories in the agricultural press and agricultural chemical magazines (8). This new technology needed testing on a larger scale. Since aerial application was indicated, we sought and obtained the excellent cooperation of Dale Hiatt of Winamac, Ind. The 1973 work was under the direction of Ken Collins, a graduate student. Since aerial spray volume was about one-fifth of what we had been using in our small experimental plots, we decided to double and quadruple what we thought would be the best rate with surface application equipment. This was probably a mistake. The replicated tests using airplanes have many limitations. The results of these tests are shown in Table III. Again strong indications of the positive effects of DNBP were indicated.

Table III.   The Response of Corn to Aerial Application of DNBP (1973)

Rate (g/a)	Yield (bu/a)	Number of Responses
0	134	
7.5	141	12 of 15
0	143	
15	144	7 of 11

In addition to the cooperative aerial trials, Collins carried out small-plot experiments at the Purdue Agronomy Farm and at Vincennes, Ind. In all earlier tests, Premerge was used in conjunction with a non-ionic wetting agent. Was it possible that all or part of the response could be ascribed to the wetting agent? To shed some light on this question, the 1973 small-plot experiments included water only and water plus wetting agent check plots. The results were dramatic at the Purdue Agronomy Farm. Instead of enhancing grain yields, the wetting agent decreased yields 55 bu/a. When water only was applied, the yield was 165 bu/a; water plus surfactant yielded 110 bu/a. 30 gal/a of solution were used in our CO_2 plot sprayer. Without surfactant the droplets remained on the upper leaves. With surfactant they coalesced, and the solution drained into the leaf whorl. Smut spores were carried into the whorl where they

germinated and severely infected the corn. The addition of 1 g/a of
DNBP in the solution resulted in a 50-bushel increase in yield (110–160
bu). This illustrated the well-established potency of DNBP as a fungicide. Higher rates of DNBP gave small increases in yield over the water
check. Although our original question was not answered, a new facet of
the use of DNBP was illustrated. It should be pointed out that spray
volumes of 3–5 gal by air and 15–20 gal by ground would not cause
solutions to run off into the leaf whorls.

In addition to our work several agricultural chemical dealers and
large farmers tested DNBP. Their results confirmed our earlier findings.
During 1973 we continued to explore possibilities for making this new
use of an old chemical available to farmers. Some type of label was
needed for this new use of an old chemical. In all experiments Dow
Chemical Co.'s Premerge (alkoamine salt of the ethanol–2-propanol series
of Dinoseb) was the product used. Since they were not interested in
obtaining a label, we explored two other avenues—the IR-4 procedure
for state institutions to obtain an EPA label and the securing of an
Indiana experimental label from the Indiana State Pesticide Administrator. Dow Chemical Co. cooperated in both efforts. Because of the
apparent lengthy procedure and large backlog of applications, the IR-4
process was given up, and the Indiana experimental label was sought.

DNBP residue analysis on 20 samples of corn grain were made by
the Indiana Pesticide Residue Laboratory. No detectable residues were
found. These results plus our experimental data and a label prepared
with the assistance and cooperation of Dow Chemical (they, of course,
had to accept the liability associated with the label) were submitted to
Mr. Hutton, State Pesticide Administrator, on May 5, 1974. On May 13,
1974 an Indiana experimental label was issued. It is conservatively
estimated that approximately 40,000 a of corn were treated in Indiana
in 1974.

Additional testing was done in 1974 by the extension service. On
seven fields, four treated and four untreated strips across the farmers'
fields were established. At six locations the DNBP was applied with
high-clearance field sprayers. At the seventh a helicopter was used.
L. Hermann not only applied six of the treatments but checked yields
(9). The results are shown in Table IV.

Because of a wet spring and an early frost, yields were low at two
locations. The consistency of the DNBP effects again added additional
confidence that the increases of about 5% were real. The consistency of
the decreases in barreness, the increases in ear number per acre, and
the weight of individual ears supported earlier observations and helped
to indicate the mode of biostimulation. Several seed-corn producers used
DNBP experimentally in 1974. Their reports indicate good efficacy but

Table IV.   Responses of Corn to 4.2 g/a of DNBP Supplied in Dow
Chemical Co.'s Premerge in Seven Replicated Field Tests (1974)

Site	Treatment (g/a)	Barren (%)	Ears (1000/a)	Grain Yield (lb/ear)	(bu/a)
Farm					
Pinney	0.0	7.38	20.0	0.368	134
Purdue	4.2	5.39	20.8	0.373	142
Davis	0.0	22.52	14.8	0.296	86
Purdue	4.2	20.52	16.2	0.305	91
Agronomy	0.0	12.90	20.0	—	70
Farm	4.2	14.45	19.6	—	71
County					
Benton	0.0	6.31	15.2	0.391	109
	4.2	4.10	15.8	0.407	118
Knox	0.0	6.80	16.3	0.384	114
	4.2	6.40	16.5	0.405	122
Pulaski	0.0	17.84	18.0	0.208	67
	4.2	14.99	18.6	0.226	75
Decatur	0.0	2.43	20.0	0.464	171
	4.2	1.69	20.0	0.471	173
Average	0.2	10.98	17.9	0.352	107.3
	4.2	9.65	18.2	0.365	113.1
Difference		1.23	0.3	0.013	5.8
Percent change		−11.3	+1.7	+3.7	+5.5

emphasized that varietal differences in response were significant. In some
cases pollination problems were diminished by treatment, but with other
lines the problem could be intensified. Extensive use must therefore rest
on adequate experimental experience.

In December of 1974 the Helena Chemical Co. contacted me concerning their interest in preparing a special formulation of DNBP for
exclusive use as a biostimulant. Needless to say, their interest was most
gratifying. After appropriate discussions and corporate decisions, Surge
was born—a formulation of DNBP (alkanolamine salts of the ethanol–2-
propanol series of Dinoseb, 2-sec-butyl–4,6-dinitrophenol), wetting agent,
and antifoaming agent. One pint of Surge, renamed Spark in 1976, as a
broadcast overall spray, treated one acre of corn. Applications for national EPA and state labels were initiated in early 1975. Only state
labels were obtained. Dow Chemical Co. also applied for labels in many

cornbelt states, giving growers two sources of DNBP for use on corn. Although there was little commercial advertising, farmer interest was strong because of stories in the agricultural press and the interest of aerial applicators. For them it was a new business with a tremendous growth potential. The aerial applicators not only advertised but also encouraged additional field testing. As a result of their efforts it is conservatively estimated that probably one-half million acres of corn were treated in the United States, of which one-quarter million acres of corn were treated in Indiana. On a national basis, nearly 1% of the corn acreage was treated, which in many respects is a remarkable accomplishment.

The 1975 growing season was unique in that corn seldom before had passed through the period of unemerged tassel elongation (the prime time for DNBP application) as rapidly as it did in 1975. Because of this rapid growth, often the biostimulant was applied too late to maximize returns. In spite of this, users reported many excellent responses.

1975 Research Plots

Several agronomy departments in the cornbelt and other states tested DNBP in 1975 with varying results. Many of these are reported in a paper by Regan (*10*). Of the results, those of R. Johnson of the University of Illinois are most puzzling. At Dixon Springs, DNBP was tested on four different hybrids at three rates. A statistically significant decrease in yield was obtained. Why this occurred is still unanswered. These factors may be involved in the answer: 40 gal of spray solution were used; runoff into the whorl would occur with possible attendant problems; and the corn was sprayed 13 days before tasseling—slightly late. Also difficult to explain was the increased tendency toward barreness with simultaneous increases in double ears. These are diametrically opposed effects resulting from the same chemical. In other tests, when the chemical was applied at the proper time, yield increases of 3–10% resulted.

Other Crops

Bruinsma's work with DNOC on rye suggests that DNBP might be effective on other members of the grass family. Results on wheat (unpublished data) at Illinois and Indiana have been sufficiently encouraging to continue studies in Indiana. Earlier applications need to be tested.

Current studies are under way on sorghum. Indiana experiments include time $\times$ rate $\times$ variety experiments. Other experiments are underway at the University of Southern Illinois and by H. D. Fuehring in New Mexico. Again, unpublished results are sufficiently encouraging to continue research.

Two or three experiments have been carried out at Purdue on soybeans. The results—no grain yield increase—did not justify further study. On this scanty data the DNBP yield enhancement effect is possibly limited to members of the grass family.

Mode of Action

At the cellular and subcellular level the mode of action is well documented in journals and textbooks (11). How these modes of action would enhance economic yield is open to speculation. Recent analysis of total nitrogen in corn grain by Ohlrogge in Indiana and Fuehring in New Mexico strongly suggests nitrogen metabolism involvement because total nitrogen concentration in the grain was increased with DNBP application (unpublished data). Hatley (4) originally suggested that grain yield is enhanced by lengthening the grain-filling period through earlier pollination of the grain. Although this observation is not always reported, it is fairly consistently observed. Physiological maturity appears to be unaffected as indicated by black layer development in the grain. I have ocacsionally observed drier grain at harvest time although this certainly is not a universal observation.

The Future

The use of PGRs on agronomic crops is in its infancy whereas in the horticultural field they have been used extensively for many years. Field crop usage has been limited to CCC (2-chloroethyltrimethylammonium chloride) on wheat in Europe and the short-lived commercialization of TIBA (Regim 8) on soybeans in the United States. Why has this occurred when more than 20 years ago A. C. Leopold predicted that the impact of PGRs on agriculture might be equivalent to the introduction of the mechanical harvester (12)? Agronomists in general have expressed little active interest or involvement in PGR development. This posture is changing with the increasing number of physiologists being found in university agronomy departments. In part, the agronomist disinterest may be caused by the fact that major agricultural chemical companies treat the state agronomists as a testing agency for experimental compounds that are nearly ready for the marketplace. Such testing is hardly appropriate for a research institution but fits well into agricultural extension programs. I would suggest that the agronomist can add much to a development program if he is a full cooperative partner. Many of the land-grant universities, through their research foundations and experiment stations, are fully able to accommodate all types of joint development programs. In a hungry world I believe that PGRs will soon make a meaningful contribution.

Literature Cited

1. Sherbeck, T. G., Hatley, O. E., Oplinger, E. S., Ramirez, R., Ohlrogge, A. J., "Chemotherapeutic Compounds in Fertilizer Bands" (1974) *Trans 10th Congress, Int. Soil Sci. Soc.,* Moscow, Russia, Com. IV, 203-208.
2. Crafts, A. S., "Toxicity of Certain Herbicides in Soils," *Hilgardia* (1945) **16,** 469.
3. Bruinsma, J., "The Variability of the Yield-Increasing Effect of a Spray with DNOC (4,6-Dinitro–*o*-cresol) in Winter Rye (*Secale Cereale* L.)" (1963) Publication No. 28, Plant Physiological Research Center, Wageninen, Netherlands.
4. Hatley, O. E., "The Response of Corn (*Zea mays* L.) and Soybeans (*Glycine max* L.) Merrill to Seed and Foliar Applications of Growth Regulating Compounds," Ph.D. thesis, Purdue University (1970).
5. Winter, S. R., "Leaf Angle Source-Sink and Detasseling Studies with Corn and Studies on Increased Reproductive Growth Period of Soybeans," Ph.D. thesis, Purdue University (1971).
6. "Herbicide Handbook of the Weed Society of America," Weed Science Society of America, Champaign, Ill., 1974, p. 156.
7. Bauer, M. E., Sherbeck, T. G., Ohlrogge, A. J., "Effects of Rate, Time, and Method of Application of TIBA on Soybean Production," *Agron. J.* (1969) **61,** 604.
8. Hatley, O. E., Herrman, L., Ohlrogge, A. J., "Foliar Applications of 'Premerge' Weed Killer Increases Corn Grain Yield," *Down Earth* (1974) **29:** 1, 4.
9. Herrman, L. G., "The Response of Corn (*Zea mays* L.) and Soybeans (*Glycine max* L.) Merrill to Seed and Foliar Applications of Growth Regulating Compounds," M.S. thesis, Purdue University (1974).
10. Regan, J. B., "The Use of Premerge at Low Rates as a Yield Stimulant in Corn," Status report to the Weed Science Society of America, Milwaukee, 1975, in press.
11. Ashton, F. M., Crafts, A. S., "Mode of Action of Herbicides," Wiley Interscience, New York, 1973.
12. Leopold, A. C., "Auxin and Plant Growth," University of California Press, Berkeley, Calif., 1955, p. 3.

RECEIVED September 22, 1976.

INDEX

INDEX

The text of this book is set in 10 point Caledonia with two points of leading. The chapter numerals are set in 30 point Garamond; the chapter titles are set in 18 point Garamond Bold.

The book is printed in offset on White Decision Opaque, 60 pound. The cover is Joanna Book Binding blue linen.

Jacket design is by Alan Kahan.
Editing and production by Kevin C. Sullivan.

The book was composed by Service Comopsition Co., Baltimore, Md., printed and bound by The Maple Press Co., York, Pa.